Introduction to Aquaculture

Introduction to Aquaculture

Brendan Marshall

SYRAWOOD
PUBLISHING HOUSE
New York

Published by Syrawood Publishing House,
750 Third Avenue, 9th Floor,
New York, NY 10017, USA
www.syrawoodpublishinghouse.com

Introduction to Aquaculture
Brendan Marshall

International Standard Book Number: 978-1-68286-817-1 (Hardback)

Cataloging-in-Publication Data

Introduction to aquaculture / Brendan Marshall.
     p. cm.
Includes bibliographical references and index.
ISBN 978-1-68286-817-1
1. Aquaculture. 2. Mariculture. 3. Agriculture. I. Marshall, Brendan.
SH135 .I58 2019
639.8--dc23

TABLE OF CONTENTS

PREFACE

Aquaculture is the farming of molluscs, crustaceans, fish, algae and aquatic plants. It involves the cultivation of economically important species, creating interventions in the rearing process for enhanced production, besides regular feeding and stocking. Shrimp, oyster and fish farming, algaculture, mariculture and the cultivation of ornamental fish are some common forms of aquaculture. Aquaponics and integrated multi-trophic aquaculture are techniques that integrate aquatic plant and fish farming. This textbook is a valuable compilation of topics, ranging from the basic to the most complex theories and principles of the aquaculture. It provides comprehensive insights into this field. Coherent flow of topics, student-friendly language and extensive use of examples make this book an invaluable source of knowledge.

A detailed account of the significant topics covered in this book is provided below:

Chapter 1- Aquaculture is the cultivation of freshwater and saltwater aquatic organisms like fish, molluscs, crustaceans, alage or aquatic plants under controlled conditions. This chapter has been carefully written to provide an introduction to aquaculture and discusses the essential topics of fish farming, mariculture, algaculture, sustainable aquaculture, etc.

Chapter 2- In aquaculture, an artificial channel called a raceway is constructed to grow aquatic organisms. It consists of canals or rectangular basins that are made of concrete and has an inlet and an outlet. Fish and other organisms are also cultivated in static water systems such as static freshwater ponds. This is known as static water aquaculture. A detailed analysis of static and flowing water aquaculture has been provided in this chapter, which includes various topics related to static freshwater ponds, water flow in a raceway, raceway systems operation, etc.

Chapter 3- Fish are commercially raised in tanks or fish ponds in fish farming. An important entity involved in raising and harvesting fish is a fishery from which wild fish, farmed fish, fresh water or saltwater fish may be captured. Fisheries can be freshwater or saltwater. This chapter elucidates the fundamental aspects of intensive and extensive aquaculture, fish farming types, issues, indoor fish farming, etc.

Chapter 4- Mariculture is an area of aquaculture, which involves the farming of marine organisms in the open ocean, in enclosed section of the ocean, tanks, raceways or ponds that are filled with seawater. An understanding of mariculture requires a study of the cultivation of algae, shellfish, etc. which have been extensively discussed in this chapter.

Chapter 5- Unregulated aquaculture practices can be environmentally damaging. Some of the concerns are the adverse effects of antibiotics, improper waste handling, farmed and wild species competition, invasive plant and animal species introduction, etc. This is an important chapter, which analyzes the environmental impact of aquaculture such as pollution, habitat destruction, disease transfer, etc.

Chapter 6- Any system that integrates aquaculture and hydroponics in a symbiotic environment is known as aquaponics. In this system, the water from an aquaculture system is sent to a hydroponic system and the by-products are broken into nitrites and nitrates by nitrifying bacteria which are then used by plants. The filtered water then recirculates back to the aquaculture. All such important aspects of hydroponic subsystem, deep water culture, fish stocking, etc. have been covered in this chapter.

It gives me an immense pleasure to thank our entire team for their efforts. Finally in the end, I would like to thank my family and colleagues who have been a great source of inspiration and support.

Brendan Marshall

Chapter 1

Aquaculture: An Introduction

Aquaculture is the cultivation of freshwater and saltwater aquatic organisms like fish, molluscs, crustaceans, alage or aquatic plants under controlled conditions. This chapter has been carefully written to provide an introduction to aquaculture and discusses the essential topics of fish farming, mariculture, algaculture, sustainable aquaculture, etc.

Aquaculture is the process of rearing, breading and harvesting of aquatic species, both animals and plants, in controlled aquatic environments like the oceans, lakes, rivers, ponds and streams. It serves different purposes including; food production, restoration of threatened and endangered species populations, wild stock population enhancement, building of aquariums, and fish cultures and habitat restoration. Here are the various types of aquaculture as well as their importance.

Types of Aquaculture

Mariculture

Mariculture is aquaculture that involves the use of sea water. It can either be done next to an ocean, with a sectioned off part of the ocean or in ponds separate from the ocean, but containing sea water all the same. The organisms bred here range from molluscs to sea food options like prawn and other shellfish, and even seaweed.

Growing plants like seaweed are also part of mariculture. These sea plant and animal species find many uses in manufacturing industries such as in cosmetic and jewellery where collagen from seaweed is used to make facial creams. Pearls are picked from mollusc and made into fashion items.

Fish Farming

Fish farming is the most common type of aquaculture. It involves the selective breeding of fish, either in fresh water or sea water, with the purpose of producing a food source for consumption. Fish farming is highly exploited as it allows for the production of cheap source of protein.

Furthermore, fish farming is easier to do than other kinds of farming as fish are not care intensive, only requiring food and proper water conditions as well temperatures. The process is also less land intensive as the size of ponds required to grow some fish species such as tilapia is much smaller than the space required to grow the same amount of protein from beef cattle.

Algaculture

Algaculture is a type of aquaculture involving the cultivation of algae. Algae are microbial organisms that share animal and plant characteristics in that they are motile sometimes like other microbes but they also contain chloroplasts that make them green and allow them to photosynthesise just like green plants. However, for economic feasiblity, they have to be grown and harvested in large numbers. Algae are finding many applications in today's markets. Exxon mobile has been making strides in developing them as a new source of energy.

Intergrated Multitrophic Aquaculture

IMTA is an advanced system of aquaculture where different trophic levels are mixed into the system to provide different nutritional needs for each other. Notably, it is an efficient system because it tries to emulate the ecological system that exists in the natural habitat.

The IMTA makes use of these intertrophic transfer of resources to ensure maximum resource utilization by using the waste of larger organisms as food sources for the smaller ones. The practice ensures the nutrients are recycled, meaning the process is less wasteful and produces more products.

Benefits of Aquaculture

Economic Benefits

Alternative Food Source

Fish and other seafood are good sources of protein. They also have more nutritional value like the addition of natural oils into the diet such as omega 3 fatty acids. Also since it offers white meat, it is better for the blood in reducing cholesterol levels as opposed to beef's red meat. Fish is also easier to keep compared to other meat producing animals as they are able to convert more feed into protein. Therefore, its overall conversion of pound of food to pound of protein makes it cheaper to rear fish as they use the food more efficiently.

Alternative Fuel Source

Algae are slowly being developed into alternative fuel sources by having them produce fuels that can replace the contemporary fossil fuels. Algae produce lipids that if harvested can be burn as an alternative fuel source whose only by products would be water when burnt.

Such a breakthrough could ease the dependency of the world on drilled fossil fuels as well as reduce the price of energy by having it grown instead of drilling petroleum. Moreover, algae fuel is cleaner and farmable source of energy, which means it can revolutionize the energy sector and create a more stable economy that avoids the boom-bust nature of oil and replaces it with a more abundant fuel source.

Increase Jobs in the Market

Aquaculture increases the number of possible jobs in the market as it provides both new products for a market and create job opportunities because of the labor required to maintain the pools and harvest the organisms grown. The increase in jobs is mostly realized in third world countries as aquaculture provides both a food source and an extra source of income to supplement those who live in these regions.

Aquaculture also saves fishermen time as they do not have to spend their days at sea fishing. It allows them free time to pursue other economic activities like engaging in alternative businesses. This increase in entrepreneurship provides more hiring possibilities and more jobs.

Reduce Sea Food Trade Deficit

The sea food trade in America is mainly based on trade from Asia and Europe, with most of it being imported. The resultant balance places a trade deficit on the nation. Aquaculture would provide a means for the reduction of this deficit at a lower opportunity cost as local production would mean that the sea food would be fresher. It would also be cheaper due to reduce transport costs.

Environmental Benefits

Creates Barrier Against Pollution with Mollusc and Sea Weed

Molluscs are filter feeders while seaweed acts a lot like the grass of the sea. Both these organisms sift the water that flows through them as brought in by the current and clean the water. This provides a buffer region that protects the rest of the sea from pollution from the land, specifically from activities that disturb the sea bed and raise dust.

Also, the economic benefits of molluscs and sea weed can create more pressure from governments to protect their habitats as they serve an economic importance. The financial benefits realised provides incentive for the government to protect the seas in order to protect sea food revenue.

Reduces Fishing Pressure on Wild Stock

The practice of aquaculture allow for alternative sources of food instead of fishing the same species in their natural habitats. Population numbers of some wild stocks of some species are in danger of being depleted due to overfishing.

Aquaculture provides an alternative by allowing farmers to breed those same species in captivity and allow the wild populations to revitalize. The incentive of less labor for more gains pushes fishers to convert to fish farmers and make even more profit that before. It also allows them control of the supply of the fish in the market giving them the ability to create surplus stock or reduce their production to reap the best profits available.

Importance of Aquaculture

Sustainable use of Sea Resources

Aquaculture provides alternatives for fishing from the sea. Increase in demand for food sources and increase in globalization has led to increase in fishing. Yet, this has led fishermen to become selfish and overfish the desired or high-demand species. Through aquaculture, it provides both an alternative and opportunity for wild stocks to replenish overtime.

Conservation of Biodiversity

Aquacultures also protect biodiversity by reducing the fishing activities on wild stock in their eco-systems. By providing alternatives to fishing, there is reduced attack on the wild populations of the various species in the sea. Reduced action of fishing saves the diversity of the aquatic ecosystem from extinction due to overfishing.

Increased Efficiency, more Resources for Less Effort

Fish convert feed into body protein more efficiently than cattle or chicken production. It is much more efficient meaning that the fish companies make more food for less feed. Such an efficiency means that less food and energy is used to produce food, meaning that the production process is cheaper as well. It saves resources and even allows for more food to be produced leading to secure reserves and less stress on the environment.

Reduced Environmental Disturbance

By increasing aquaculture, fish farming in specific, there is a reduced need for the fishing of the wild stock. As an outcome, it puts less stress on the ecosystem and equally reduces human inter-ference. Actions of motor boats and other human influences such as the removal of viable breeding adult fish are all stresses put on the aquatic ecosystems and their discontinuation allows the eco-system to flourish and find their natural balance.

Fish Farming

Fish farming is also known as Pisciculture. It is the business of raising fish in large quantity, usu-ally for food. Fish farming is the principal form of aquaculture.

Systems of Fish Farming

Cage System

Cage system of fish farming utilises existing water resources but encloses the fish in a cage or basket which allows water to pass freely between the fish and the pond. The method is also called "off-shore cultivation".

Advantages

- Many types of waters can be used (rivers, lakes, filled quarries, etc.),

- Many types of fish can be raised:

- Fish farming can co-exist with sport fishing and other water uses.

Disadvantages

- Concerns of disease, poaching, poor water quality, etc., lead some to believe that in general, pond systems are easier to manage and simpler to start.

- Past occurrences of cage-failures leading to escapes, have raised concern regarding the culture of non-native fish species in dam or open-water cages.

- Even though the cage-industry has made numerous technological advances in cage construction in recent years, storms will always make the concern for escapes valid.

Irrigation Ditch or Pond System of Fish Farming

This involves using ponds or irrigation ditches to raise fish. Using this method, one can store one's water allotment in ponds or ditches, usually lined with bentonite clay. In small systems the fish are often fed commercial fish food, and their waste products can help fertilise the fields. In larger ponds, the pond grows water plants and algae as fish food.

Control of water quality is crucial. Fertilizing, clarifying and pH control of the water can increase yields substantially, as long as eutrophication is prevented and oxygen levels stay high. Yields can be low if the fish grow ill from electrolyte stress.

Raceway Fish Fa'rming

A raceway, also known as a flow-through system, is an artificial channel used in aquaculture to culture aquatic organisms. Raceway systems are among the earliest methods used for inland aqua-culture. A raceway usually consists of rectangular basins or canals constructed of concrete and equipped with an inlet and outlet. A continuous water flow-through is maintained to provide the required level of water quality, which allows animals to be cultured at higher densities within the raceway.

Freshwater species such as trout, catfish and tilapia are commonly cultured in raceways. Raceways are also used for some marine species which need a constant water flow, such as juvenile salmon, brackish water sea bass and sea bream and marine invertebrates such as abalone.

Tank System of Fish Farming

Tank system of fish farming is another artificial channel used in the business of fish farming. With a properly designed inlet system, the oxygen rich water will provide just the right current for good muscle tone while distributing new water evenly throughout the tank.

Mariculture

Mariculture is a branch of aquaculture in which aquatic species are raised within the marine environment. Over a generation ago, Jacques Cousteau shared his vision of how controlled production of aquatic species via mariculture must replace hunting of the sea for long-term sustainability and food security. The need for protecting the ocean's remaining populations is even greater today. Fortunately, there have been tremendous advances in mariculture. Oceans, estuaries, and even man-made marine environments are being used to farm shrimp, prawns, crabs, clams, mussels, oysters, abalone, sea urchins, sea cucumber, kelp, and a large variety of finfish. However, most production of marine species occurs near shore and even on-shore. There is promise that technologies and permits will soon allow mariculture to move far off-shore into the open ocean, which would open vast areas for expansion and move farms further away from more sensitive coastal environments.

Algaculture

Algaculture is the commercial cultivation of algae. Algae are simple green plants that grow in water. Their green color means they produce their own food using photosynthesis, just like grass, trees and corn. Algae come in two main forms. Macroalgae are seaweeds. Kelp grows to more than 180 feet (55 meters) long in the ocean. Nori is the variety you'll find wrapped around your sushi. Microalgae are tiny, single-celled plants that float in the water, each one visible only through a microscope.

Algaculture is nothing new. Seaweed was first cultivated in Japan at least 1,500 years ago and algae production is still a big business there. Dulse has long been eaten in the British Isles and the microalgae spirulina were harvested by the Aztecs of 16th-century Mexico. In addition to providing human food, seaweeds have been used for fertilizers. They provide the food thickener carrageen and other gelling agents and stabilizers that show up in everything from soup to toothpaste. Worldwide, algae production is a $6 billion business.

Algae, shown here floating on the top of a pond, may look humble, but have the potential to help change the energy industry -- if only we find efficient ways to cultivate them

Today, algae are attracting new interest and research investment because of their potential to provide energy and combat environmental threats. Part of the organic mass of algae takes the form of oil, which can be squeezed out and converted to biodiesel fuel. Algae beat land plants hands down in the efficiency with which they produce oil. Some varieties of algae yield oil that can be refined into gasoline and even jet fuel. The carbohydrate portion of the plants can be fermented for ethanol production.

Algae can convert waste carbon dioxide, a greenhouse gas that pours from smokestacks, to usable products. They can help clean dirty water, converting pollutants to biomass. They have additional uses in pharmaceuticals and cosmetics.

There are three basic systems for cultivating algae, each with its advantages and disadvantages:

1. Open pond: The simplest and cheapest way to grow algae is in large, shallow ponds. The water is often divided into concentric lanes or raceways, with paddlewheels to move the algae mixture in a circle. This helps bring algae to the surface, where they're exposed to light, and mixes nutrients and carbon dioxide into the liquid. The open-pond method produces less algae biomass than other methods. It loses water to evaporation, so more must be added. And it allows for contamination by predators or undesirable algae.

2. Closed pond: This method is similar to an open pond, but the water is covered by a Plexiglas greenhouse. This raises the cost, but allows greater control of the process. It reduces evaporation and contamination and extends the growing season. Growers can raise algae year-round if the space is heated.

3. Biophotoreactor: A completely closed system, the biophotoreactor consists of glass or acrylic tubes where the algae are exposed to light. Pumps move the water, nutrients and algae through the tubes and storage tanks. Some reactors automatically harvest the algae when

they're ready. This approach gives growers the most control over the process and the most efficient way to produce algae biomass. But it's also the most costly to set up and operate.

All of these systems are designed for growing microalgae, the one-celled varieties that float in water. Growers usually cultivate macroalgae in the open sea. The water already contains the nutrients the algae need and provides good growing conditions. The traditional method was simply to harvest wild seaweed, and this is still done in coastal areas around the world.

With increased demand, growers began to cultivate seaweed. For some varieties, such as kelp, spores are attached to ropes that are then anchored in the ocean and the seaweed is allowed to grow. Other types grow from pieces of seaweed that are fixed to nets or deposited in pools.

Integrated Multitrophic Aquaculture

Integrated multi-trophic aquaculture (IMTA) is the farming, in proximity, of species from different trophic levels and with complementary ecosystem functions in a way that allows one species' uneaten feed and wastes, nutrients and by-products to be recaptured and converted into fertilizer, feed and energy for the other crops, and to take advantage of synergistic interactions among species while biomitigation takes place.

Farmers combine the cultivation of fed species such as finfish or shrimp with extractive species, such as seaweeds and aquatic plants that recapture inorganic dissolved nutrients, and shellfish and other invertebrates that recapture organic particulate nutrients for their growth.

The aim is to ecologically engineer aquaculture systems for increased environmental sustainability; economic stability through improved output, lower costs, product diversification, risk reduction and job creation; and societal acceptability.

IMTA is based on a very simple principle. "The solution to nutrification is not dilution, but extraction and conversion through diversification," which is another way of expressing the principle of conservation of mass, as formulated by Antoine- Laurent de Lavoisier in 1789. "Nothing is created, nothing is lost, everything is transformed," he said.

What is important is that the appropriate organisms to be co-cultured are chosen at multiple trophic levels based on their complementary functions in the ecosystem, as well as their economic value or potential. Integration should be understood as cultivation in proximity, not considering absolute distances but connectivity in terms of ecosystemic functionalities.

The IMTA concept is extremely flexible. It is the central/overarching theme on which many variations can be developed. IMTA can be applied to open-water or land-based systems (sometimes called aquaponics), marine or freshwater systems, and temperate or tropical systems.

In the minds of those who created the acronym IMTA, it was never conceived as only the cultivation of salmon, kelps, blue mussels and other invertebrates in temperate waters and within a few hundred meters. This is only one of the variations, and the IMTA concept can be extended to very large ecosystems.

The scope of IMTA can cover a whole global aquaculture advocate March/April 2013 19 range of operations, from integrated agriculture aquaculture and integrated fisheries aquaculture to partitioned aquaculture, integrated periurban aquaculture and integrated food and renewable energy parks. All should be considered variations on the central IMTA theme.

There is no ultimate IMTA system to feed the world. Different climatic, environmental, biological, physical, chemical, economic, historical, societal, political and governance conditions, prevailing in the parts of the world where they operate, can lead to different choices in the design of the best-suited IMTA systems, but all of them are based on the same principles of the IMTA concept.

Techniques of Aquaculture

A number of aquaculture practices are used world-wide in three types of environment (freshwater, brackishwater, and marine) for a great variety of culture organisms. Freshwater aquaculture is carried out either in fish ponds, fish pens, fish cages or, on a limited scale, in rice paddies. Brackishwater aquaculture is done mainly in fish ponds located in coastal areas. Marine culture employs either fish cages or substrates for molluscs and seaweeds such as stakes, ropes, and rafts.

Culture systems range from extensive to intensive depending on the stocking density of the culture organisms, the level of inputs, and the degree of management. In countries where government priority is directed toward increased fish production from aquaculture to help meet domestic demand, either as a result of the lack of access to large waterbodies (e.g., Nepal, Central African Republic) or the over-exploitation of marine or inland fisheries (e.g., Thailand, Zambia), aquaculture practices are almost exclusively oriented toward production for domestic consumption.

These practices include:

(i)   Freshwater pond culture;

(ii) Rice-fish culture or integrated fish farming;

(iii) Brackishwater finfish culture;

(iv) Mariculture involving extensive culture and producing fish/shellfish (e.g., oysters, mussels, cockles) which are sold in rural and urban markets at relatively low prices.

Table: Aquaculture production systems and practices, by region

Region	Major Culture Species	Major Culture Systems	Major Culture Practices	Scope for Future Development/ Needs for Further Expansion
Asia	At least 75 species; diverse freshwater and marine species, including high-value shrimps, molluscs, seaweeds, with carps and seaweeds dominating production	Traditional extensive to intensive	- Fish ponds - Fish pens and fish cages - Floating rafts, lines, and stakes for molluscs and seaweeds	Development of culture-based fisheries in inland lakes, rivers, floodplains, and permanent and temporary reservoirs and barrages
				Resource enhancement programmes integrated with environmental management

Pacific	Mussels and oysters, red seaweeds	Intensive/semi-intensive to extensive	- Hanging lines for mussels and pearl oysters	Production of high-value species for select markets;
			- Offshore cages for salmon	Small-scale aquaculture for local markets;
			- Pond culture for shrimps, tilapia, catfish, milkfish	Improved management of fishery resources, particularly reef fisheries
			- Freshwater pens for crayfish	
Latin America	50 species of fish, crustaceans, and molluscs, including freshwater fish and marine shrimps in South America and molluscs in Central America	Extensive to semi-intensive and Intensive	- Offshore cage farming of Pacific and Atlantic salmon - Ocean ranching in Southern Ocean - Semi-intensive farming of marine shrimp in coastal ponds and extensive farming of freshwater fish in ponds	Production of species for export and marine shrimp and salmon
Africa	>26 freshwater fish; the most important being tilapia and common carp, molluscs and oysters also	Mainly extensive, rural-based, integrated with poultry and animal husbandry, rice-fish farming; some intensive in raceways and floating cages	- Fish pond culture for freshwater fish - Raceways and floating cages for marine species	Increased emphasis on higher value catfishes for urban markets, on marine species of fish and crustaceans for select national market and export
				Culture-based fisheries in lakes and reservoirs
				Development of coastal lagoons which are almost totally unexploited
Mediterranean	>50 individual species, mostly freshwater and brackishwater fishes - most important being salmonids and carps; oysters and mussels	Well-diversified modern practices, with highly technical and intensive systems in developing countries and semi-intensive and extensive elsewhere	- Fish pond - Fish cages - Ocean ranching	Production of high-value species of tourism and export Integrated coastal zone management

Caribbean	About 16 species of tilapias, carps, marine shrimp and, freshwater prawns, oysters and seaweeds		- Floating cages in reservoirs	Priority is for aquaculture production for local markets
			- Fish pond farming in freshwater	
			- Culture-based fisheries in reservoirs	
			- Rope production of molluscs	

Extensive systems use low stocking densities (e.g., 5 000-10 000 shrimp post larvae (PL)/ha/crop) and no supplemental feeding, although fertilization may be done to stimulate the growth and production of natural food in the water. Water change is effected through tidal means, i.e., new water is let in only during high tide and the pond can be drained only at low tide. The ponds used for extensive culture are usually large (more than two ha) and may be shallow and not fully cleared of tree stumps. Production is generally low at less than 1 t/ha/y.

Semi-intensive systems use densities higher than extensive systems (e.g., 50 000-100 000 shrimp PL/ha/crop) and use supplementary feeding. Intensive culture uses very high densities of culture organism (e.g., 200 000-300 000 shrimp PL/ha/crop) and is totally dependent on artificial, formulated feeds. Both systems use small pond compartments of up to one ha in size for ease of management.

Semi-intensive and intensive culture systems are managed by the application of inputs (mainly feeds, fertilizers, lime, and pesticides) and the manipulation of the environment primarily by way of water management through the use of pumps and aerators. Feeding of the stock is done at regular intervals during the day. In intensive shrimp culture, the computed daily feed ration is given in equal doses from as low as three to as high as six times a day. Water change is also effected on a daily basis, with approximately 10-15% of the water in the pond replenished by the entry of new water in semi-intensive shrimp ponds.

Semi-intensive and intensive culture systems are therefore more labour-intensive than extensive systems which need little attention, and are costlier to set up and operate, not to mention the fact that they also carry higher risks of mortalities resulting from disease, poor management, and/or force majeure (e.g., from anoxia due to non-functioning aerators during times of power failure).

Production is of course much higher (for example, ranging from a minimum of 1.5 t/ha/crop from semi-intensive shrimp culture to a high of 10 t/ha/crop from intensive shrimp culture). Financial returns are therefore much more attractive than those from extensive culture, although studies have shown that the return on investment (ROI) from semi-intensive culture is better than from intensive culture due to the high cost of inputs (largely fry and feeds) used in intensive culture.

Fish Pond Culture

Pond culture, or the breeding and rearing of fish in natural or artificial basins, is the earliest form of aquaculture with its origins dating back to the era of the Yin Dynasty. Over the years, the practice has spread to almost all parts of the world and is used for a wide variety of culture organisms in freshwater, brackishwater, and marine environments. It is carried out mostly using stagnant waters but can also be used in running waters especially in highland sites with flowing water.

Table: Summary of comparative features among the three main culture systems

Parameter	Extensive	Semi-Intensive	Intensive
Species Used	Monoculture or Polyculture	Monoculture	Monoculture
Stocking Rate	Moderate	Higher than extensive culture	Maximum
Engineering Design and Layout	May or may not be well laid-out	With provisions for effective water management	Very well engineered system with pumps and aerators to control water quality and quantity
	Very big ponds	Manageable-sized units (up to 2 ha each)	Small ponds, usually 0.5-1 ha each
	Ponds may or may not be fully cleaned	Fully cleaned ponds	Fully cleaned ponds
Fertilizer	Used to enhance natural productivity	Used regularly with lime	Not used
Pesticides	Not used	Used regularly for prohylaxis	Used regularly for prophylaxis
Food and Feeding Regimen	None	Regular feeding of high quality feeds	Full feeding of high-quality feeds
		Depending on stocking density used, formulated feeds may be used partially or totally	
Cropping Frequency (crops/y)	2	2.5	2.5
Quality of Product	Good quality	Good quality	Good quality
	Culture species dominant but extraneous species may occur	Confined to culture species	Confined to culture species
	Variable sizes	Uniform sizes	Uniform sizes

Running water fish culture involves growing the fingerlings to marketable size in earthen ponds using water from rivers, irrigation canals, or plain rain water. The system approximates intensive culture in that it involves the application of rapid water changes and the heavy stocking of the cultured species. The continuously flowing water is advantageous for fish culture as it supplies abundant dissolved oxygen and flushes away waste products and unconsumed feeds.

Culture Species

Commonly raised species in freshwater ponds are the carps, tilapia, catfish, snakehead, eel, trout, goldfish, gouramy, trout, pike, tench, salmonids, palaemonids, and the giant freshwater prawn Macrobrachium. In brackishwater ponds, common species include milkfish (Chanos chanos), mullet (Mugil sp.) and the different penaeid shrimps (Penaeus monodon, P. orientalis, P. merguiensis, P. penicillatus, P. semisulcatus, P. japonicus, and M. ensis). The more popular species for culture in marine ponds are the sea bass, grouper, red sea bream, yellowtail, rabbitfish, and marine shrimps.

In Asia, where the bulk of world production from aquaculture emanates, fish ponds are mostly freshwater or brackishwater, and rarely marine. In China and most of the Indian sub-continent, pond culture is traditionally dominated by freshwater species, mainly the carps, usually in polyculture and/or integrated with animal husbandry. In Southeast Asia, fish ponds are predominantly brackishwater, with milkfish and penaeid shrimps grown either in polyculture or in monoculture.

Recently in Latin America and the Caribbean, brackishwater pond culture of penaeid shrimps has expanded rapidly, as it has in some parts of Asia.

In Africa, the tilapias and carps dominate aquaculture production. Controlled breeding is also carried out in ponds with goldfish, trout, Bagrus and, to a lesser extent, Lates niloticus,Heterotis niloticus, and Clarias lazera. Ten species of molluscs belonging to four genera (Crassostrea, Mytilus, Venerupis and Pinctada) are cultured. Crustacean culture has yet to be developed on a significant scale.

Site Selection

Proper site selection is recognized as the first step guaranteeing the eventual success of any aquaculture project and forms the basis for the design, layout, and management of the project. For fish ponds, especially those to be used for coastal/brackishwater aquaculture of high-value species like shrimps, site selection is critical and should be given utmost attention.

Adisukresno, Hechanova, and Jamandre and Rabanal listed the following guidelines for the selection of a suitable site for coastal fish ponds:

(i) Soil Quality: preferably, clay-loam, or sandy-clay for water retention and suitability for diking; alkaline pH (7 and above) to prevent problems that result from acid-sulphate soils (e.g., poor fertilizer response; low natural food production and slow growth of culture species; probable fish kills).

(ii) Land elevation and tidal characteristics; preferably with average elevation that can be watered by ordinary high tides and drained by ordinary low tides; tidal fluctuation preferably moderate at 2-3 m. (Sites where tidal fluctuation is large, say 4 m, are not suitable because they would require very large, expensive dikes to prevent flooding during high tide. On the other hand, areas with slight tidal fluctuation, say 1 m or less, could not be drained or filled properly.)

(iii) Vegetation; preferably without big tree stumps and thick vegetation which entail large expense for clearing; areas near river banks and those at coastal shores exposed to wave action require a buffer zone with substantial growths of mangrove. (The presence of Avicennia indicates productive soil; nipa and trees with high tannin content indicate low pH.)

(iv) Water supply and quality: with steady supply of both fresh and brackish water in adequate quantities throughout the year; water supply should be pollution-free and with a pH of 7.8-8.5.

(v) Accessibility: preferably readily accessible by land/water transport; close to sources of inputs such as fry, feeds, fertilizers, and markets, fish ports, processing plants, and ice plants; and linked by communication facilities to major centres.

(vi)   Availability of manpower for construction and operation.

Pond Layout

The layout of the pond system depends on the species for culture and on the size and shape of the area, which in turn determines the number and sizes of ponds and the position of the water canals and gates. A fish farm is considered properly planned if all the water control structures, canals, and the different pond compartments mutually complement each other. A complete fish farm has nursery and grow-out ponds and, in some instances, transition ponds for intermediate-sized fish/ shrimp, all of which are properly proportioned and positioned within.

Milkfish culture in brackishwater ponds in the Philippines follows the traditional practice of providing for nursery, transition, and rearing operations. In some cases, formation ponds are used for additional growth or stunting of fingerlings prior to stocking in rearing ponds. The nursery ponds comprise about 1-4% of the total production area while the transition and formation ponds constitute about 6-9% of total area.

It has been suggested that a similar progressive culture scheme be adopted for shrimp pond culture when no supplementary feeding is practised. For growing to a medium size, a two-stage progression composed of a nursery pond (NP) and a rearing pond (RP) is adequate; for growing to larger sizes, a three-stage progression composed of nursery, transition, and rearing ponds is recommended.

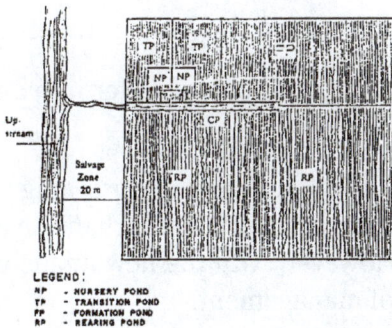

Figure: Layout of conventional pond system

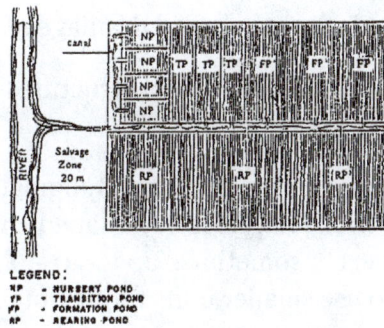

Figure: Modular pond system for milkfish culture

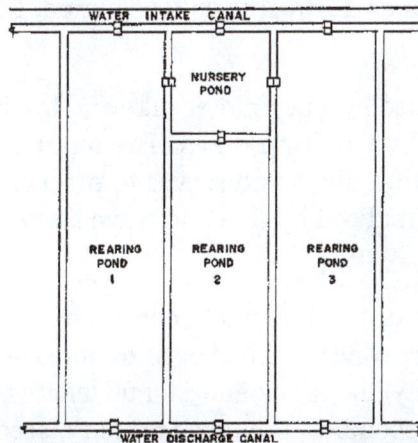

Figure: Pond layout with one nursery pond and three rearing ponds

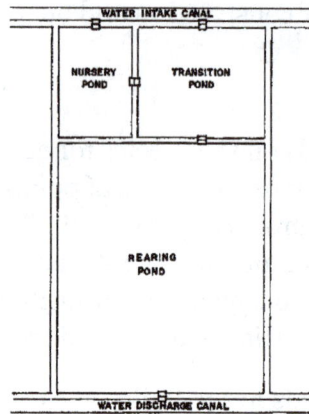

Figure: Pond layout with one nursery pond, one transition pond, and one rearing pond

In general, however, shrimp monoculture uses direct stocking of post larvae in rearing ponds and therefore requires only one type of pond with separate inlets and outlets for better circulation and aeration.

Design of Pond Facilities

A fish pond system consists of the following basic components:

(i)   Pond compartments enclosed by dikes;

(ii) Canals for supply and drainage of water to and from the pond compartments; and

(iii) Gates or water control structures to regulate entry and exit of water into and from the pond compartments.

Pond compartments are usually rectangular in shape although in Indonesia, running water ponds are generally triangular, raceway-shaped, or oval. They vary in size from less than a hectare to several hectares each; sometimes up to 20-50 ha in size. However, with the new intensive methods, the trend is to use smaller units for flexibility and ease of management.

The elevation of the rearing pond bottom for milkfish is usually such that only a maximum of 40 cm of water can be held in the ponds during the culture period. For new shrimp ponds, the minimum water depth is 1 m.

The entire pond system is enclosed by a perimeter dike and the individual pond compartments are separated from each other by partition dikes. The outer perimeter dike is usually wider and higher than the inner partition dikes and serves to protect the entire fish pond area from flooding and destruction brought about by tide and wave action. The inner dikes are narrower and shorter.

The design of the dikes depends primarily on soil characteristics. Dikes are usually earthen although intensive shrimp ponds are concrete-lined or brick-lined as in Taiwan (PC). The side slopes are designed for structural stability, the ratio of horizontal length to height ranging from 1:1 to 1:3. The height and width of dikes depend on the type (primary, secondary, or tertiary), tide conditions, flood level, pond water depth, soil shrinkage, and freeboard.

The following slopes are recommended for dikes built with good clay soil:

- 2:1 when dike height is above 4.26 m and exposed to wave action;

- 1:1 when dike height is less than 4.26 m and tidal range is greater than 1 m; and

- 1:2 when tidal range is 1 m or less, and dike height is less than 1m.

The dike crown should not be less than 0.5 m and the main dike surrounding the farm should be 0.5 m above the highest dike or flood level recorded in the locality.

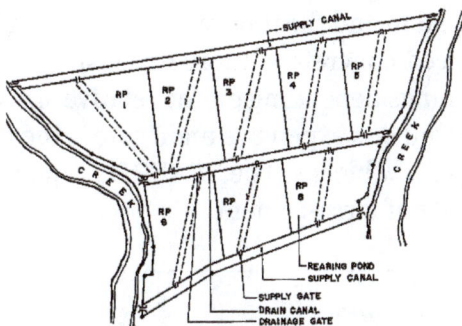

Figure: Pond layout showing shrimp pond compartments, canals, and gates

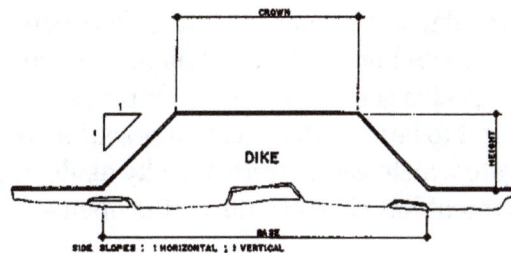

Figure: Typical cross sections of dikes

Water conveyance structures (canals/channels) supply new water into the pond and drain out old water. They also provide the facility for holding and harvesting of fish and of serving as waterways for transporting farm supplies. Traditional milkfish ponds usually have only one canal that is used for both supply and drainage. Shrimp ponds have separate supply and drainage canals. Canals which are to be used for harvesting should be 30 cm below the level of the pond bottom to allow draining of pond water.

Having separate water intake and discharge canals in a pond complex brings about the following advantages:

(i)   Better filling and non-contamination of pond by discharge from other ponds,

(ii) Greatly reduced possibility of spread of disease,

(iii) Maintenance of constant head in intake canal thus reducing water loss through leaks/ seepages in pond dikes and consequently reducing leaching of acids into the ponds from dikes with acid-sulphate soils,

(iv) Absence of conflict of usage between farmers,

(v)   Better water exchange for individual ponds, and

(vi) Possibility of effecting flow-through systems.

The width of the canals depends on the amount of water they must carry. The following should be taken into account when designing canals:

(i) Volume of water to be held in the ponds.

(ii) Time requirement for filling or draining the pond.

(iii)   Amount of rainfall which must be carried off in a given period of time.

(iv)   Elevation of canal bottom in relation to tide.

(v)   Other uses like transportation, harvesting of milkfish, and holding of broodstock.

Diversion canals are constructed where there is much runoff from adjoining areas, to prevent sudden salinity changes and the possible entry of polluted, pesticide-loaded water and/or of silted water into the pond complex.

The entry and exit of water into ponds through the canals is regulated or controlled by gates. Main gates regulate the exchange of water between the pond system and the tidal stream or sea, and may be constructed of reinforced concrete or wood. Reinforced concrete is more expensive but lasts longer. Such a gate has one or multiple (2, 3, 4, etc.) openings depending on the relative size of the pond unit to be served. A recent innovation for a smaller and less expensive main gate is the monk-type gate which uses culverts usually made of concrete hollow blocks. The SEAFDEC Aquaculture Department has also introduced the open sluice gate made of ferro-cement

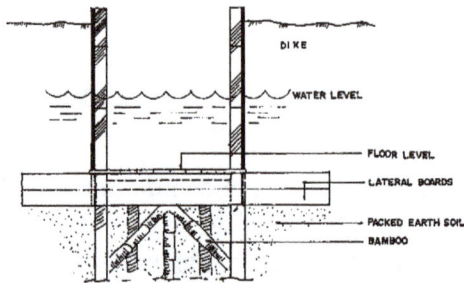

Figure: Diagram of wooden gate

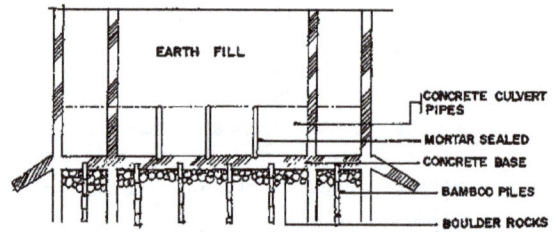

Figure: Use of culvert pipes as secondary gates

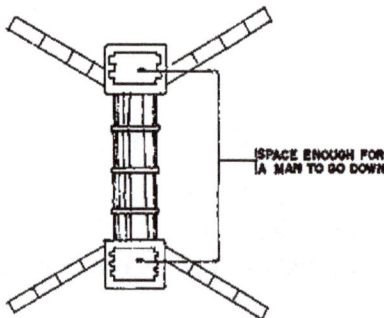

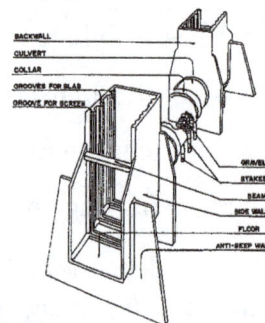

Figure: Use of culvert pipes as secondary gates

Figure: Use of culvert pipes as secondary gates

Figure: Ferro-cement culvert developed at the SEAFDEC Aquaculture Department

Secondary gates, which regulate water exchange between the ponds and the canals, are usually made of wood. Pipes or culverts can also be used for smaller ponds such as nursery or fry ponds and transition ponds for milkfish culture. Secondary gates are now usually located toward one end of the narrower side of the pond compartment to give good turbulence and circulation during the filling and draining.

Shrimp ponds are provided with separate supply and drainage gates to effect flow-through water management and facilitate water exchange through supply and drainage canals. Inlet and outlet gates are best located at opposite corners of the same pond across which a diagonal trench, about 5-10 m wide and 0.3-0.5 m deep, extending from inlet to outlet gates is recommended for convenient draining of water.

Gates should be located where they are not exposed to strong weather forces and where water of good quality can be allowed to enter the fish pond system. Proper gate location can also serve to aerate the pond water and promote water circulation.

During the construction of gates for shrimp ponds a number of requirements should be kept in mind, and the gates should:

   (i)   Be durable, water-tight, and made of locally available materials;

   (ii) Have adequate capacity for the amount of water to be taken in or drained;

   (iii) Allow water to be taken in or discharged at the bottom;

   (iv) Have provisions for draining pond surface water;

   (v)   Have gate bottom elevation that permits complete draining of pond water;

   (vi) Have slots or grooves for the placement of outside and inside screens to prevent undesirable species from entering the pond and the shrimps from leaving the pond;

   (vii) Have place for net installation for harvesting; and

   (viii) Be easy to operate.

Pond Management

Pond management techniques for finfish and shrimp culture, while varying slightly depending on the specific biological requirements of the culture organism, the type of culture system, and the culture environment (freshwater, brackishwater, and marine), are similar in that they involve the following basic activities:

   (i)   Pond preparation/conditioning.

   (ii) Stocking.

   (iii) Feeding and/or fertilization (depending on the culture system used).

   (iv) Water management.

(v)   Pond maintenance.

(vi) Harvesting.

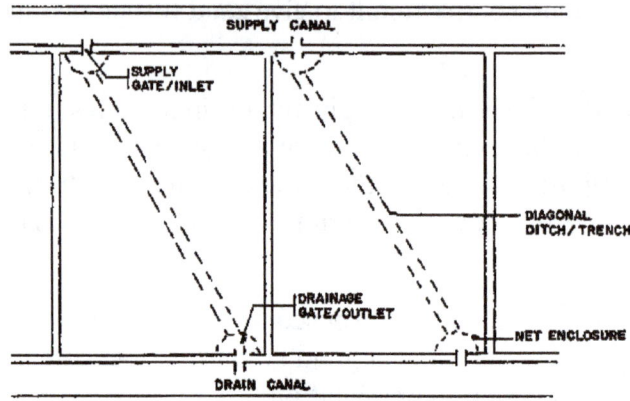

Figure: Layout of improved shrimp pond showing diagonal trench extending from inlet to outlet

Variations would consist mainly of differences in application rates of fertilizers, lime, pesticides, and feeds; stocking rates and sizes of stocking material; rate of water change; and harvesting techniques.

Extensively managed systems generally require the least management, with no supplemental feeding and minimal water exchange on account of the low stocking density used. On the other hand, intensively managed ponds require full artificial feeding and substantial water management to ensure optimum culture conditions for the species being reared.

Pond Preparation

Ponds are totally drained and the pond bottoms dried prior to the application of pesticides. Tobacco dust, derris root/rotenone powder, teaseed cake/powder, or Gusathion-A are used to eliminate predators and/or wild species that may eventually compete with the cultured organisms for food and space. Teaseed cake is perhaps the best fish poison to use in brackishwater ponds to selectively kill unwanted fish without damaging the shrimps and without affecting rotifers and copepods which are feed for shrimps. On the other hand, rotenone is most effective in fresh water and works better in low-salinity water.

Ponds with acid-sulphate soils are repeatedly dried and flushed, i.e., filled and drained to remove the acids formed by pyrite oxidation. Agricultural lime is then applied to correct soil pH and bring it up to at least 6.5. Brackishwater ponds are usually treated by spreading 1.5 t of agricultural lime per ha, followed by another 1.5 t worked into the soil.

To stimulate and maintain the growth of natural plankton, organic (e.g., chicken manure) or inorganic fertilizer (e.g., urea, ammonium phosphate) are applied to the pond bottom. After fertilizer application, water is let in to a depth of about 20-40 cm and gradually increased to 1 m a week after fertilization. Intensively managed ponds or ponds where artificial feeding shall be given, do not need to be fertilized. Extensive ponds need regular fertilization during the culture period to maintain the growth of natural food. Semi-intensive ponds may use a mix of fertilization and supplementary feeding.

Table: Variations in pond management techniques commonly used for different species

Species	Stocking Rate	Fertilization	Feed Type	Rate of Water Change	Pesticides/Predator Control	
					Type	Application Rate
Milkfish (Chanos chanos)	2 000-5 000/ha	16-20-0 at 50 kg/ha; 45-0-0 at 15 kg/ha; chicken manure at 0.5 t/ha twice weekly	Rice bran and trash fish as supplemental feed	Once every two weeks at high tide	Lime; ammonium sulfate	1 t/ha 10 g/m²
Tilapia (O. niloticus; O. mossambicus)	5 000-20 000/ha	Chicken manure at 500 kg/ha; Inorganic fertilizers at 50 kg/ha	Rice bran, fish meal, ipil-ipil leaf meal			
Catfish (Clariasbotrachus and monocephalus)	60-300/m²		9 parts trash fish and 1 part rice by-products	When necessary		
Penaeids	From as low as 15 000 to as high as 300 000/ha	Chicken manure at 1-2 t/ha followed by inorganic fertilizer at 75-150 kg/ha mono-ammonium phosphate (16-20-0) and 25-50 kg/ha of urea (46-0-0)	Supplemental feed of rice bran with trash fish, mussels, and clam meat; artificial/formulated diets with 40% CP.	20-30% once every week or every two weeks for low density ponds; 5-20% daily for semi-intensive to intensive ponds		

Stocking

After the pond is prepared, fish fingerlings or shrimp post larvae are stocked at the appropriate density depending on the culture strategy, size of pond, and the size of fingerlings, among others.

The fingerlings are properly acclimated and conditioned prior to stocking and weak or diseased fish eliminated. Stocking is usually done in the early morning or late afternoon.

Feeding

Fish/shrimp grown in semi-intensive and intensive culture ponds are given supplementary and full artificial feeds, respectively, the former to augment the natural food in the pond, the latter to totally replace the natural organisms in the water as a source of nutrition.

A wide variety of feed ingredients is used to prepare supplemental/artificial feeds. The simplest fish feeds are prepared at the pond site using locally available raw materials like rice or corn bran, copra meal, and rice mill sweepings as sources of carbohydrates. These are usually mixed with animal protein like trash fish/fish meal, shrimp heads, and snail meat. Supplemental feeds for tilapia are prepared using 80% rice bran and 20% fish meal. Those for shrimps in improved extensive culture (low-density stocking but given dietary supplements for increased growth/production) usually include fresh raw materials like snail/mussel/clam meat or carabao hide and other slaughterhouse leftovers.

Commercial feed preparations are also available now in a wide range of brandnames, mostly for semi-intensive and intensive shrimp culture. (Taiwan (PC), Japan, and the USA are the top producers of commercial fish/shrimp feeds.) These commercial diets consist of a number of ingredients like fish meal, blood meal, bone meat, and shrimp head meal (to serve as attractant for the shrimp), together with vitamin and mineral premix and carbohydrate sources like rice/corn bran or wheat. The crude protein (CP) content of these shrimp feeds is generally not lower than 30% to satisfy the high animal protein requirement of shrimps, actually estimated to be about 40% during the earlier stages of growth.

Commercial feeds usually come in various formulations to match the protein requirement of the culture organism, which as a rule, decreases with age. Thus, fish/shrimp feeds come in different forms as starter, grower, and finisher, with starter feeds having the highest CP content of about 40% and finisher feeds having the lowest CP content of about 20%. Starter feeds are usually given on the first month of culture, finisher feeds on the last month, and grower feeds in between.

Some shrimp culturists prefer not to give artificial feeds during the first two weeks of culture when the newly stocked post larvae can subsist on the plankton available in the water.

The feeding rate is computed as a percentage of the estimated animal biomass in the pond, with higher rations given when the animals are small and gradually decreasing as they become bigger. The daily feeding rate usually starts at 5% and 10-15% of estimated biomass of fish and shrimps, respectively, and decreases to a low of 2% and 5%, for fish and shrimps, respectively, toward harvest.

The daily feed rations are given in equal portions during the course of a day. Freshwater fish like tilapia are usually fed twice a day - early morning and late afternoon. Penaeid shrimps are fed more frequently, from three to four to as often as six to seven times a day.

Feeds are broadcast into the water and/or supplied on feeding trays. In semi-intensive and intensive shrimp ponds, small feeding boats are used by caretakers who go around the pond distributing the feed by broadcasting. At certain points along the periphery of the pond, feeding trays are submerged into the water after known quantities of feed are put on the surface, to supply feed to the shrimps in the pond as well as to monitor feed consumption and shrimp growth. The feeding tray is lifted two to three hours after the feed was supplied to check how much of it has been consumed and to see if the shrimps are healthy and feeding. Empty feeding trays may indicate that the quantity given is inadequate and may have to be increased. Conversely, full or slightly touched trays indicate excessive feed quantities and/or sluggish shrimps. The feeding ration is subsequently adjusted accordingly to optimize feed utilization.

By monitoring the feeding tray, one can get a good indication of the sizes and quantity of shrimps present in the pond without a need for cast-netting or actual sampling, since shrimps are invariably found on the tray when it is lifted out of the water.

Water Management

Water in the pond is kept at certain levels for optimal fish growth. In general, a pond water depth of 1 meter is considered best for culture of tilapia, carps, and shrimps; traditional milkfish ponds can do with just 40-60 cm of water.

Pond water is not just maintained at a certain depth; its quality must also be kept high to ensure optimal growth of the culture organism. This is particularly important in semi-intensive and intensive culture systems where large amounts of metabolites are continuously excreted into the pond and where excess, unconsumed feeds add to the bottom load and serve to pollute the water.

To prevent the deterioration of the pond environment, pond water is continuously freshened by the entry of new water from the river or water source (through the supply canal) while old water is drained through the outlet/drainage gate and through the drainage canal into the sea or river.

A flow-through system of water management that allows the simultaneous entry and exit of water into and out of the pond is essential in any high-density culture system. This is effected by the provision of separate inlets and outlets for all the ponds, each inlet regulating the flow of water from the supply canal to the pond and each outlet controlling the discharge of water out of the pond into the drainage canal. Both the supply and drain gates are so designed as to bring water into and drain water out of the lower levels of the pond, where water quality tends to get poorer faster as a result of the accumulation of wastes and their subsequent decomposition.

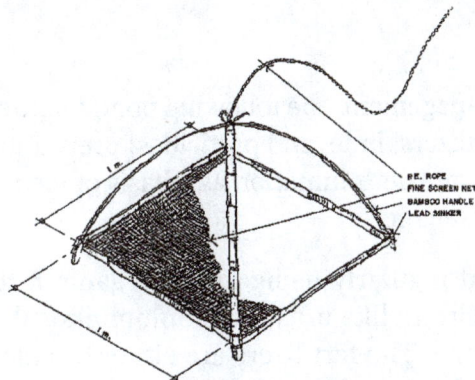

Figure: Feeding tray

The regular replenishment of pond water, independent of natural tidal fluctuations, is made possible by the use of pumps which draw water from the source even at low tide. Although there is no hard-and-fast rule as to the rate of water change necessary for medium- to high density aquaculture, semi-intensive culture systems usually change water at the rate of 10% daily for an equivalent total replacement of water every ten days or three times per month. Intensively managed ponds require greater water exchange in view of the much higher organic load on the pond bottom, especially toward the latter part of the culture cycle when the animals excrete more wastes.

Intensive ponds/tanks usually need to provide for aeration facilities/equipment to prevent anoxia that may lead to mass mortalities. Oxygen depletion in high-density ponds results not only from the faster rate of utilization of dissolved oxygen for respiratory activities; it is also caused by the fast rate of decomposition at the pond bottom by aerobic or oxygen-consuming micro-organisms.

Paddlewheels or other types of aerators are thus provided in the ponds to effect the infusion/introduction of greater quantities of oxygen into the water and prevent fish/shrimp mortalities. The aerators are usually operated at regular/periodic intervals for certain fixed durations during the day but especially in the early morning hours when the concentration of dissolved oxygen is known to be lowest (as a result of the absence of photosynthetic, oxygen-producing activity in the pond). Toward the end of the culture period when oxygen demand is highest, aeration may have to be provided continuously and not just sporadically as could be done during the initial stages of rearing. At that time too, water pumps usually need to be run for longer periods to effect greater water exchange.

Pond water is also regularly sampled and measurements taken of basic/essential parameters, particularly dissolved oxygen, pH, and salinity. This is important for the purpose of determining the need for corrective/remedial action to bring water quality to optimum levels and obtain good yields.

Dissolved oxygen levels are kept, as much as possible, above 5 ppm by pumping and aeration. Problems of acidity are corrected by liming. Salinity is an important parameter for penaeid culture and has to be maintained within a range of 15-25 ppt for best results. During summer months, high-salinity water can be diluted by mixing with fresh water from springs or deep wells.

Pond Maintenance

Fertilization

Aside from feeds and water management, the following pond maintenance procedures are carried out: regular application of fertilizers, lime, and pesticides; prevention of entry of predators; monitoring of the stock for growth rate determination as a basis of feeds and water management; and regular pond upkeep and maintenance.

Extensive ponds are fertilized regularly using either organic fertilizers like chicken, cow, or pig manure, or inorganic fertilizers like urea, ammonium phosphate, or both, to maintain the plankton population in the pond. The fertilizers are either broadcast over the pond water surface or kept in sacks suspended from poles staked at certain portions along the pond periphery. Semi-intensive and intensive culture systems do not require fertilization since they are not natural food-based, except for those which grow plankton-feeders like milkfish whose diet is largely algae dependent.

Liming

In addition to fertilization, ponds also need to be given regular doses of lime to maintain water pH at alkaline or near-alkaline levels (preferably not lower than six). Agricultural lime is broadcast over the pond and applied on the sides of the dikes to correct soil and water acidity.

Elimination of Pests and Predators

Unwanted and predatory species which may have survived the application of pesticides during pond preparation or which were able to enter the pond through the gate screens or through cracks in the dikes, are eliminated by the application of pesticides, preferably organic, into the pond.

Crabs, which are a serious problem in shrimp ponds because they are carnivorous and cause damage to the pond dikes, are not usually affected by known pesticides and are therefore best eliminated by the use of crab traps situated in the pond.

It is also important that the gates are properly screened and the screens kept whole, to prevent the entry of small unwanted fish into the pond. Double screens are usually installed at the main intake to ensure that pests and predators are prevented from entering the pond system.

Stock Monitoring

The culture organisms are monitored closely and regularly to determine their rate of growth and the general condition of the stock. They are regularly sampled for length-weight measurements as a basis for determining/estimating their biomass in the pond and therefore their daily feed rations, as well as for making projections on harvest schedules and procurement of pond inputs.

In the first few months of culture, the feeding tray is a good tool for stock monitoring. As the organisms grow in size, cast-netting is used as a sampling tool, with those caught in the throw of the cast net providing an indication as to sizes and weights of stock. Based on the sampled weights and the daily feed consumption, it is possible to predict the available biomass (i.e., stock surviving after initial mortalities) and make projections on volume of harvest. For this purpose, it is essential that accurate records are kept for analysis at a later time. Data on initial size/weight and number of fry/post larvae stocked, average body weight at each sampling, and feed consumption on a daily basis, are important to have on file.

Regular Upkeep and Maintenance of Facilities

The pond dike and gates are checked regularly for cracks that could lead to seepages and losses of stock. The dikes are best planted with grass or vegetative cover to prevent erosion. The gates and other support infrastructure are properly maintained for efficient operation.

Harvesting

Marketable-size fish/shrimps are harvested at the end of the culture period by draining the pond and using harvesting nets to catch the fish or shrimps. The latter are harvested with a bag-net attached to the sluice gate as water is drained out of the pond at low tide. Tilapia are harvested using seine nets after the pond water is drained to half-level the night before.

Harvest of milkfish takes advantage of their behaviour of swimming against the current. The method, known in the Philippines as "pasulang" or "pasubang" involves draining 85-90% of the pond water during low tide and allowing in the water at the incoming high tide so that the fish swim against the current through the tertiary gate and into the catching pond, whose gate is closed once a large number of fish is impounded. The fish in the catching pond are then harvested by seining and the rest hand-picked.

Integrated Fish Farming

In a number of countries in Asia (e.g., China, Nepal, Thailand, Malaysia, Indonesia) and in some parts of Africa, freshwater fish culture is integrated with the farming of crops, mainly rice, vegetables and animals (usually pigs, ducks, and chickens). This leads to greater overall efficiency of the farming system as wastes/by-products or one component are used as inputs in another. For example, poultry or pig manure can be used to fertilize the fish pond and the vegetable garden and the waste vegetables can be fed to the fish and the pigs.

In Africa, fish culture in rice fields and in combination with pig and duck rearing, is not too widely practised but has significant potential. Reported fish yields ranged from 2 000-4 000 kg/ha/y with ducks, 8 500-8 900 kg/ha/y with pigs, and 3 600-4 900 kg/ha/y with poultry in Gabon. It has also been proven economically viable since it involves minimal investment. Its spread has, however, been constrained by the widespread use of pesticides in many countries.

Pen and Cage Culture

Pen and cage culture involve the rearing of fish within fixed or floating net enclosures supported by frameworks made of bamboo, wood, or metal, and set in sheltered, shallow portions of lakes, bays, rivers, and estuaries.

Compared to fish pond culture with its 4 000-year tradition, fish pen/cage culture is of more recent origin. Cage culture seems to have developed independently in at least two countries - in Kampuchea where fishermen in and around the Great Lake region would keep Clarias spp. and other commercial fishes in bamboo or rattan cages and baskets; and in Indonesia where bamboo cages have been used to grow Leptobarbus hoeveni fry as early as 1922. Since then, cage culture has spread throughout the world to more than 35 countries in Europe, Asia, Africa, and the Americas.

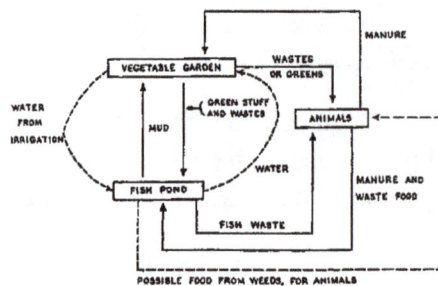

Figure: Interrelationships among the various components of an integrated fishfarming system

Pen culture is said to have originated in the Inland Sea area of Japan in the early 1920s, adopted by the People's Republic of China in the 1950s for rearing carps in freshwater lakes, and introduced to culture milkfish in the shallow, freshwater, eutrophic Laguna de Bay in the Philippines in the

1970s. From there it has been successfully extended for the culture of tilapia and carps. Its development and adoption as a popular technology has not been widespread, though, perhaps because of its site-specific requirements like its suitability mainly in shallow lentic environments. At present, it is commercially practised only in the Philippines, Indonesia, and China.

The wider popularity of cage culture as compared to pen culture may be due to its greater flexibility in terms of siting the structures. For example, cages may be installed in bays, lagoons, straits, and open coasts as long as they are protected from strong monsoonal winds and rough seas. Floating cages can also be set up in deep lakes and reservoirs, and in rivers and canal systems, and even in deep mining pools which could not be used otherwise for culture due to harvesting difficulties.

In general, however, both pen and cage culture have expanded rapidly, especially over the past two decades vis-a-vis the decreasing availability of land-based resources for fish culture and an increasing awareness of their merits over traditional pond culture, such as:

(i)   Their applicability in different types of open water bodies like coastal waters, protected coves and bays, lakes, rivers, and reservoirs;

(ii) Their high productivity (of as much as 10-20 times that of ponds Of comparative sizes) with minimal inputs and at lower costs to develop and operate; and

(iii) The Greater socio-economic opportunities they provide to low-income families in the rural areas, particularly those displaced by the reduction of fish catches in over-exploited coastal, municipal waters, because they require comparatively low capital outlay and use simple technology.

Yields from pen and cage culture are generally high, with or without supplemental feeding depending on the natural productivity of the water body. In the Philippines, for example, the yields of milkfish from fish pens in Laguna de Bay were as high as 4 t/ha/crop (compared to a national milkfish fish pond average of 1 t/ha/y in 1980 when the productivity of the lake was very high at 1 700 mg C/m^3/hr.

In Indonesia, the cage culture of common carp in the Lido Reservoir in Cigombong gave a total production of 28 kg/m^2 at a stocking density of 6 kg/m^2. The cage culture of marine finfishes has likewise been shown to give high yields.

Culture Species

The choice of species for stocking and rearing in pens and cages is governed by much the same criteria as in species selection for pond culture, including:

(i)   fast growth in confinement;

(ii) good consumer acceptance;

(iii) high tolerance to a wide range of environmental conditions;

(iv) resistance to disease;

(v)   ready supply of fish seed for stocking; and

(vi)   ease of culture and management.

Table: Comparison of production of cage-cultured marine fish

Species		Seriola T: quinquer-adiata	Trachinotus carolinus	Polydactylus sexfilis	Epinephelus sal-moides*
Country of culture		Japan	Florida, USA	Hawaii, USA	Penang, Malaysia
Initial stocking density					
	fish/m³	10	250	50	60
	kg/m³	0.15-0.55	1.75	0.4	3.4
Rearing period (days)		225	273	300	240
Production (kg/m³)		0.85-14.45	44.7	-	41.4
Average production rate (kg/m³/day)		0.004-0.06	0.16	-	0.17
Mean size of fish					
	Initial (g)	10-50	7	9	55.7
	At harvest (g)	1 000-2 000	213.6	300	795.9
Average growth rate (g/fish/day)		4.40-8.67	0.76	0.97	3.08

*Based on existing commercial culture.

There are approximately ten species of fish which are commercially cultured in cages and pens in both temperate and tropical waters, including tilapias (S. mossambicus and S.niloticus); carps (Chinese, Indian, and common varieties); milkfish; snakeheads and catfishes; marble goby; and salmonids (rainbow trout, salmon). Marine species include mainly grouper, sea bass, mullet, snapper, and milkfish.

In the Philippines, Indonesia, and China, pen culture is limited to the following species: milkfish (Chanos chanos); tilapia; and the Chinese carps: bighead (Aristichthys nobilis), silver carp (Hypophthalmichthys molitrix), grass carp (Ctenophanyngodon idella); and common carp (Cyprinus carpio).

Other species have been suggested as possible candidates for utilization in pen/cage culture in the following three different environments :

- Freshwater

    Habitats with high natural productivity (e.g., lakes, oxbow lakes, swamps, mining pools, rivers, and reservoirs): mullets, eels, catfish, Puntius gonionotus.

    Habitats with low natural productivity: Leptobarbus, Clarias batrachus, Oxyeleotris, and Macrobrachium.

- Brackishwater

    Sea bass, mullet, siganids, sea bream, grouper, snapper, threadfin, carangids. Hilsa spp., Sparus spp., and eels.

- Marine

    Siganids, pampano, yellowtail, tuna, grouper, snapper, sea bass, sea bream, carangids, pomfret.

Site Selection

The selection of sites for fish pen/cage culture should be guided by the following basic criteria:

Table: Commercially important species in inland water cage and pen farming

Species		Countries	Climate	Type of feeding	Lotic/Lentic	Cage/Pen
Salmonids	Rainbow trout	Europe, North America, Japan, high altitude tropics (eg Colombia, Bolivia, Papua New Guinea)	Temperate	Intensive. High protein (40%)	Lentic	Floating cage
	Salmon (various species) smolts	Europe, North America, South America, Japan	Temperate	Intensive. High protein (45)	Lentic	Floating cage
Carps	Chinese carps (Silver carp, grass carp, bighead carp)	Asia, Europe, North America	Temperate-tropical	Mainly semi-intensive, although also extensive (Asia) and intensive (Europe North America)	Lotic and lentic	Cages and pens
	Indian major carps (Labeorohita)	Asia	Sub-tropical-tropical	Semi-intensive	Mainly lentic	Mainly cages
	Common carp	Asia, Europe, North America, South America	Temperature-tropical	Mainly semi-intensive, although also intensive	Mainly lentic	Mainly cages
Tilapias	(O. Mossambicus, O. niloticus, etc.)	Asia, Africa, North America, South America	Sub-tropical-tropical	Mainly semi-intensive, although also intensive	Mainly lentic	Mainly cages
Catfishes	Channel catfish	North America	Temperature-sub-tropical	Intensive	Lentic	Floating cages
	Clarias spp.	Southeast Asia, Africa	Tropical	Semi-intensive	Lotic and lentic	Floating cages
Snakeheads	Channa spp. Ophicephalus spp.	Southeast Asia	Tropical	Semi-intensive/ intensive	Lotic and lentic	Floating cages
Pangasius spp.		Southeast Asia	Tropical	Semi-intensive	Lentic	Floating cages
Milkfish		Southeast Asia	Tropical	Semi-intensive	Lentic	Pens

(i)   Protection from high winds or typhoons.

(ii) Adequate water exchange that will enable the flow of nutrient-laden water through the pens/cages.

(iii) Good water quality (high or adequate dissolved oxygen, stable pH, and low turbidity, and absence of pollution).

(iv) Firm bottom mud to allow pen framework to be driven deep into substrate for better support.

(v)   Freedom from predators and natural hazards.

(vi) Accessibility to sources of inputs, including labour and markets, and

(vii) Good peace and order condition.

The factors to be considered in selecting sites for pens and cages in freshwater, brackishwater, and marine environments are shown in table above. It is important to note that the selection of a suitable site is vital to the success of the culture system; a good site selected solves much of the management problems of pen/cage culture.

Design and Construction

Both fish pens and fish cages are built around the same basic design concept: a net enclosure supported by a rigid framework. They differ, however, in a number of respects. Firstly, a pen does not have a net bottom; the edges of its net wallings/fencings are anchored to the lake bottom/substrate by means of bamboo pegs and the lake bottom is the pen bottom. In comparison, a cage is like an inverted mosquito net with the cage bottom made of the same netting material used for its four sides.

Secondly, fish pens theoretically have no limit to their size/area while cages cannot exceed 1 000 m² in area for reasons of the quantity of material required for cage construction (due to the need for a flooring) and manageability of operation (cages have to be lifted and the fish scooped out and not harvested using nets as in pens).

Thirdly, design of the structures and methods of construction are different. Fish pens are fixed structures; fish cages may either be fixed or floating. Fish pens for milkfish culture in Laguna de Bay, Philippines consist of a nursery pen within the grow-out pen/enclosure. Cages are individual units for either seed production or grow-out; they are, however, usually installed in clusters or modules with a common framework.

Pens and cages come in various shapes and sizes and are made of different types of materials. Most pens and cages are rectangular or square although some may be circular, as in some milkfish pens in Laguna de Bay and the milkfish broodstock cages at the SEAFDEC Aquaculture Department in the Philippines, or cylindrical as those used for fish collection in Malaysian or Indonesian fresh waters. Rectangular cages are preferred for easy operation and management. Circular cages are more suitable for some species like milkfish and yellowtail but are more expensive to build.

Table: Factors to be considered in the selection of cage/pen sites

	Marine	Brackishwater	Freshwater
Protection from Elements			
Natural	Wind direction	Water current	Wind direction
	Lagoons, bays and coves offer	Erosion and	Water current
	differing	accretion	Floods
	situations	siltation	Typhoons
Artificial	Breakwaters	Deflectors	Breakwaters
Water Circulation			
Related to protection	Currents	Currents	Currents
	Tidal levels	Tidal levels	Stratification and up-welling
Net pen spacing	Well-spaced	Well-spaced	Well-spaced
Water Quality and Soil Type			
Chemical	Salinity	Salinity	Soil type
	Type of bottom	Type of bottom	pH, NH_3, BOD, hardness
		Pesticides and fertilizer run-off	Saltwater intrusion
Physical	Temperature	Temperature	Temperature
	Siltation and turbidity	Siltation and turbidity	Siltation and turbidity
	Tidal fluctuation	Tidal fluctuation	Depth fluctuation
	Topography	Topography	Texture of the substratum
		Floating objects	Topography
			Floating objects
Biological	Predators, pests and competitors	Predators, pests and competitors	Algal bloom
	Vegetation	Vegetation	Predators, pests and competitors
	Plankton bloom	Plankton and benthos	Vegetation
	Diseases and parasites	Diseases and parasites	Diseases and parasites
			Natural productivity
Pollution	Industrial pollutants	Industrial pollutants	Industrial pollutants
			Thermal pollution
	Domestic pollutants	Domestic pollutants	Agricultural pollutants
	Agricultural pollutants	Agricultural pollutants	Mine pollution
Access and Security			
Supplies	Materials	Materials	Materials
	Feeds	Feeds	Feeds
	Fingerlings	Fingerlings	Fingerlings
Markets (live and fresh sales)	Close to market	Close to market	Close to market
Labor	Availability	Availability	Availability
	Cost	Cost	Cost
Monitoring	Easy access necessary for regular monitoring visits.		
Security	Efficient precautions and security from interference of all sorts.		
Others	Frequency of navigation	Frequency of navigation	Frequency of navigation
	Property rights, policies and laws	Property rights, policies and laws	Property rights, policies and laws
	Social aspects	Social aspects	Social aspects

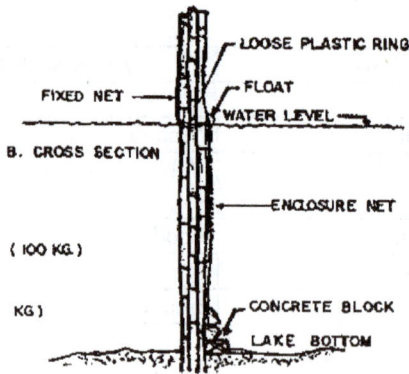

Figure: Indicative design of a fishpen wall showing how it is anchored on the lake bottom

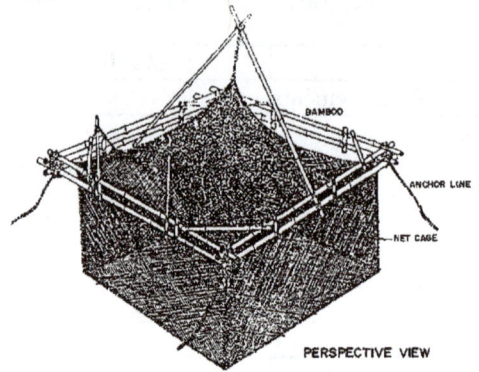

Figure: Perspective view and parts of a floating cage

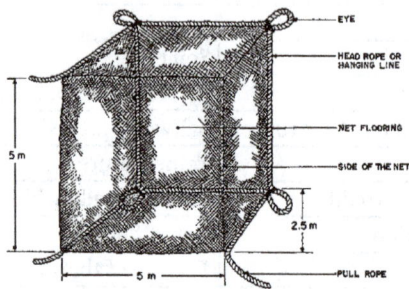

Figure: Perspective of a fishpen showing nursery pen within the grow-out enclosure

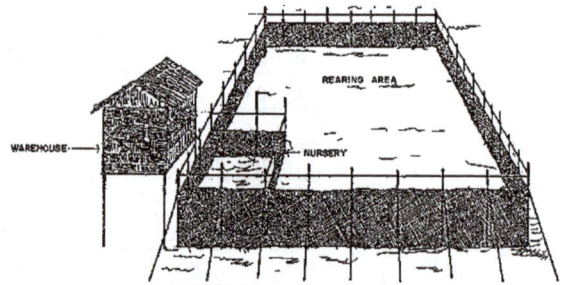

Figure: Cluster/module of fish cages

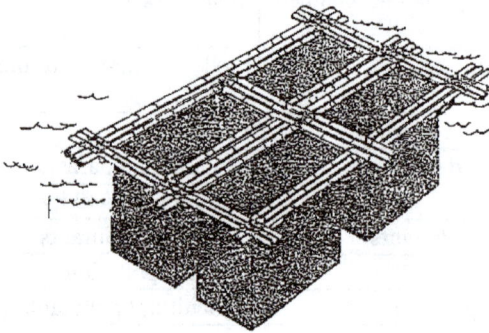

Figure: Circular milkfish broodstock cage used at the SEAFDEC Aquaculture Department

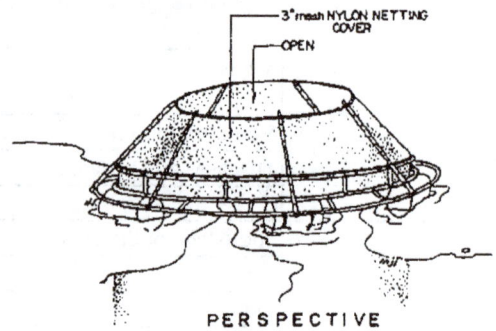

Figure: Cylindrical fish cage made of bamboo and rattan

Polyethylene and nylon monofilament twine are widely used for fabricating cages and net pens although wire mesh is used in several countries. The framework structure is generally made out of bamboo and other locally available wood. Cage floatation materials include bamboo, PVC pipes/

containers, steel or plastic drums, styrofoam, and aluminum floats. The type of anchor for floating cages varies depending on the depth of water, nature of bottom, tides, and currents. Concrete slabs of different sizes and shapes, sand bags, and iron anchors are widely used in different countries.

Pen and Cage Operation

Basic procedures involved in the management of pen and cage culture are very much like those in pond culture, starting with completion of construction and preparation of the culture facilities for stocking, rearing, and harvesting. Slight variations in specific activities exist, however, as the result of the very nature of the system. For example, it is obviously not possible to apply fertilizers, lime, and pesticides since the system has open water exchange between the inner compartment and the outside environment.

Soon after construction of the pen/cage is completed, preparations are made to procure fry/fingerlings for stocking. Milkfish pens have a nursery compartment into which milkfish fry are grown for 3-4 weeks to 12 cm long fingerlings which can be released into the grow-out compartment.

The nursery pen and the grow-out compartment are prepared for stocking by clearing the bottom of predatory fish like Megalops cyprinoides and Elops hawaiiensis. The milkfish fry/fingerlings from the nursery pen are stocked in the rearing pen at 20 000-50 000 per ha where they are cultured to marketable size.

In the Philippines, the milkfish stock in the pen is not generally given supplemental feeding except for occasional rations of bread crumbs, rice bran, broken ice cream cones, fish meal, and ipil-ipil leaf mill.

On the other hand, cage-reared fish may or may not be fed supplemental or artificial diets depending on the stocking density used and the level of technology in the country. Cage feeding trials in the Philippines showed the adequacy of a ration composed of 77% rice bran and 23% fish meal with feed conversion ratios of 2.2-2.8. Current feed practices in freshwater cage culture involve the provision of supplemental feeds using readily available ingredients like rice bran and poultry feeds. Other countries use artificial feeds based on simple diets preferably prepared in pelleted form for best results.

At the end of the culture period, the fish are harvested from pens using harvesting nets (e.g., gill nets, cast nets, seines) or from cages by lifting the cage and causing the fish to collect in one corner for scooping out using a pail.

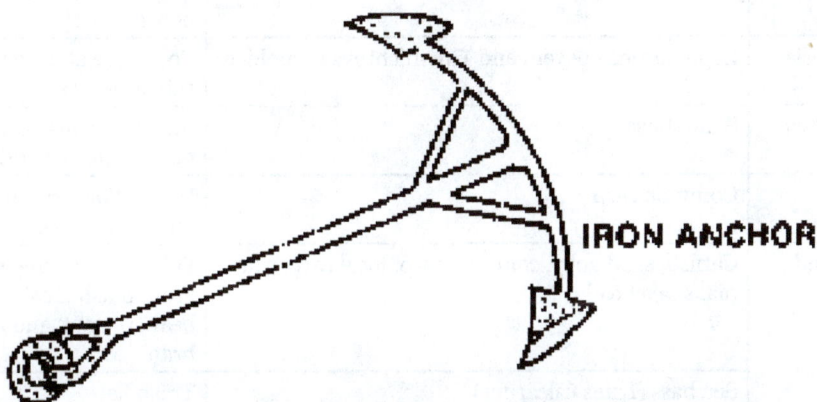

IRON ANCHOR

Figure: Types of anchor used for floating cages

Figure: Types of anchor used for floating cages

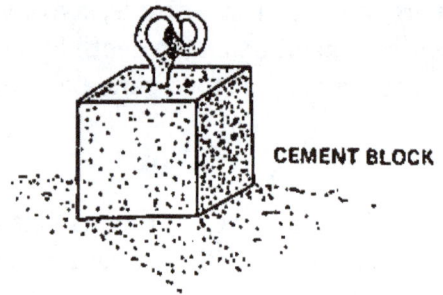

Figure: Types of anchor used for floating cages

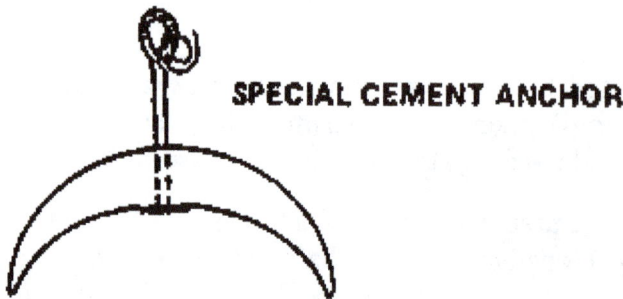

Figure: Types of anchor used for floating cages

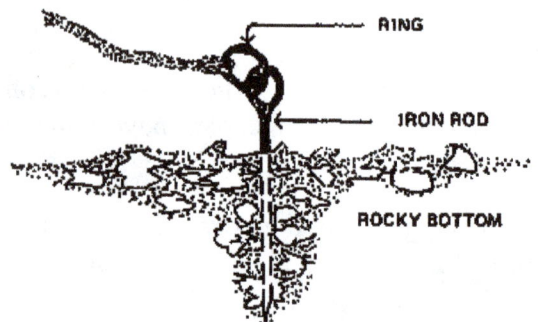

Figure: Types of anchor used for floating cages

Country	Culture Species	Feed Type
GDR	Common carp	Formulated feed/pellets, 33.7% CP
USSR	Common carp	Mixture of minced trash fish, molluscs, crayfish, and grown cereals
Hungary	Wels (Silurens glanis)	Trash fish, slaughter-house wastes, cereal grain meals
-do-	Carp polyculture (common, silver, bighead)	Pelleted common carp feed
India	Indian carp polyculture	Soya bean powder, ground nut, oil cake, rice polish (1:1.1)
Indonesia	Leptobarbus hoeveni and Thynnichthys thynoides	Coconut water, cassava, rubber leaves
Indonesia	S. niloticus	Aquatic plants (Lemna, Hydrila, Chara)
Nepal	Common carp	Wheat flour, rice bran, mustard oil cake
Thailand	Catfish, sand goby, common carp, local carp, tilapia, snakehead	Pellets consisting of ground fish meal, soy bean, peanut, and rice bran
	Sea bass (Lates calcarifer)	Trash fish

Open Water Culture

The farming of molluscs and seaweeds in open marine waters has become increasingly popular in a number of countries, especially in the Third World where it is seen as a viable alternative to municipal or artisanal fisheries or as a means of supplementary income for small-scale fishermen. Because seafarming is generally low-cost and labour-intensive and could thus involve entire coastal communities, it is particularly appropriate in areas where production from municipal fisheries has substantially declined and where, as a result, subsistence fishermen have little or no means of livelihood.

Mollusc Culture

Bivalves are widely cultured in a number of countries world-wide. In Asia and the Pacific, they represent a high quality food resource with annual production higher than from crustacean culture on a per hectare basis. In 1984, molluscs accounted for approximately 35% of the total production of coastal aquaculture in terms of gross weight in the region.

The most important species for culture in Southeast Asia are the oysters (mainly Crassostrea spp.), mussels (mainly Perna spp.), clams, cockles, and scallops.

In Japan, the most commonly cultured species include Crassostrea gigas, C. rivularis, C. nippona, C. echinata, and Ostrea denseramellosa, with C. gigas as the predominant species (Honma, 1980). In Africa, the culture of Venerupis is reported in Tunisia and Pinctada spp. in Sudan. In Mexico, the culture of the large oyster Crassostrea spp. is carried out by cooperative societies and of the mussel Mytilus edulis on floating rafts by private investors.

Oysters are widely distributed in estuaries and bays which receive some run-off from land and have somewhat lower salinity than the open sea. As they filter their food from the water, they grow best in areas with moderate to high concentrations of phytoplankton. Oysters grow best in intertidal areas where they are exposed for some minutes or a few hours during low tide. Mussels, on the other hand, cannot tolerate tidal exposure even during low tide.

The best sites for culturing molluscs are therefore those that meet their biological requirements, including the following:

(i) Seawater salinity range of 15-35 ppt.

(ii) Water depth of 1-10 m, and

(iii) Muddy bottom for mussels and hard rocky or coralline substrates for oysters.

In addition, the area for mollusc culture should be protected from strong water currents reaching three knots and should be accessible to source of seed, transport, and markets. Furthermore, the presence of local available stock in an area is a good indicator of its suitability for mollusc culture.

Countries which have successfully cultured bivalve molluscs have developed their own systems of culture which depend entirely on natural seed stock, which are either gathered from natural seed beds or collected using suitable materials for collecting seed from natural grounds.

In the Philippines, both natural and synthetic ropes have been used for spat collection. However, since natural ropes, which have been found to attract more larvae than synthetic polyethylene or polypropylene ropes, do not last long, natural fibrous materials like coconut coir are sometimes interwoven with synthetic nylon ropes to make them more attractive to the larvae.

The string seed collectors are submerged in the sea water for seed collection at the right time. They are hung on a collector rack, normally 12 strings along a distance of 1.8 m to hold about 1 000 shells. Sometimes, strings are hung separately from each other at regular intervals; at others, three or four strings are put together for hanging to prevent branches from attaching to strings when they occur in large quantities.

Three principal methods of oyster culture are used in the Philippines and Japan: (i) hanging method including rafts, longlines, simple hanging, and rocks; (ii) stake or stick method; and (iii) broadcast or sowing method.

In Japan, the earliest method used at the Hiroshima Prefecture, where oyster culture began in the 17th century, was the stick culture method. In 1927, the hanging method of culture was introduced which later developed into different variations, viz., the simple hanging method, raft method, and longline method, to suit different local conditions as culture grounds shifted from inner to outer parts of the bay to outer open seas.

The broadcast system is actually used throughout the world in places where the bottom of shallow bays is firm enough to support the materials used as collectors and for growing oysters. Oyster shells, stones, or other hard objects are scattered on the bottom in areas where setting or the attachment of oyster larvae is known to occur. The young oysters or spat are left in places attached to the collectors until they are large enough for harvest.

The stake method is usually applied in shallow areas with soft or muddy bottom, usually not more than 1 m deep during low tide. The stakes, usually bamboo trunks (whole or split), branches of mangrove trees, or concrete Y-shaped posts and other similar materials are staked on the sea bottom in rows spaced about 0.5 m apart, to serve as attachment for oyster spat.

The hanging method of oyster culture uses empty oyster shells or other material such as coconut shells as collectors. The collectors are strung on synthetic twine or heavy monofilament nylon, and placed about 10 cm apart by using bamboo tubes as spacers or by tying knots in the twine. The strings are hung from a platform or rack/tray made of bamboo or wooden splits or welded wire with wooden frame, and placed on wooden plots. Oysters detached from the collectors or those small oysters/seedlings which are separated from harvested stocks are cultured on the trays until they are big enough for the market.

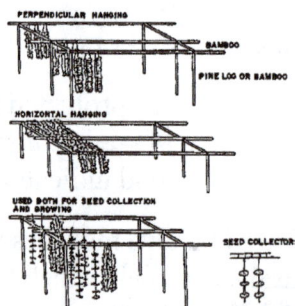

Figure: String seed collectors for mollusc spat

Harvesting procedures vary with the culture method. Oysters grown on stakes or by hanging are removed from the stakes or ropes on shore or in a boat after the stakes/ropes are lifted out of the water. Those grown by broadcasting are usually collected at low tide.

Mussel farming makes extensive use of bamboos either as stakes or as floating rafts. The stake method, similar to that for oyster culture, is the most commonly used. The mussels are harvested by divers after 6-10 months when they reach a length of 5-8 cm.

Alternatively, mussels are grown on floating rafts which have the following advantages: (i) faster growth; (ii) possibility of regular thinning and therefore higher production per unit area; (iii) possibility of transfer to other areas to prevent siltation; and (iv) ease of construction using more durable materials.

Mussels and oysters grown in waters contaminated by domestic and industrial wastes need to undergo depuration or cleansing, using artificially cleaned water or clean seawater from saltwater wells, to ensure satisfactory microbiological and chemical quality of the product.

Seaweed Farming

Seaweeds, aside from being used as food, are important sources of colloids or gels, such as agar, as well as minerals of medicinal importance such as iodine. Eucheuma, a red algae, is a valuable source of carrageenan, an important industrial compound used in stabilizing and improving the quality of a great number of products. Caulerpa lentillifera, a green algae, is economically important because it is a favourite and nutritious salad dish containing essential trace minerals such as calcium, potassium, magnesium, sodium, copper, iron and zinc. It is also known for its medicinal properties, being used as an anti-fungal agent and as a natural means for lowering blood pressure. Gracilaria, another red alga, is economically important in Taiwan (PC) for its agar extracts.

The culture of the seaweed Porphyra is believed to have started as early as between 1596 and 1614 in Hiroshima Bay utilizing pole and net devices originally installed to catch fish. At present, commercial seaweed culture is limited to five countries in East Asia, viz., Japan and Korea (which both grow mainly Porphyra, Undaria and Laminaria), China (Porphyra and Laminaria), Taiwan (PC) (Gracilaria and Porphyra), and the Philippines (Eucheuma spinosium, E. cottonii and Caulerpa lentillifera). Thirty-one species belonging to 18 genera and three divisions are presently cultured in these five countries, of which only three out of the 31 species are green algae.

In 1988, the estimated world seaweed production for use in the manufacture of carrageenan was nearly 68 000 t of dried seaweeds, of which nearly 66% was supplied by the Philippines and the rest by Indonesia, Chile and Canada. The bulk of the Philippine seaweed production consists of Eucheuma produced mainly in the southern part of the country in reef-protected coastal areas. Caulerpa is also successfully farmed in seawater ponds in Mactan, Cebu.

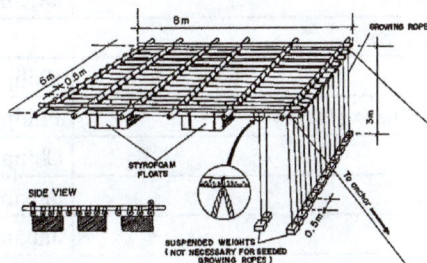

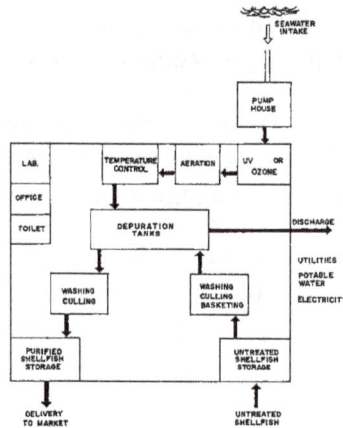

Figure: A mussel raft unit

Seaweed Groups/Species	Country Where Cultivated
A. Green Seaweeds (Chlorophyta)	
Caulerpa lentillifera J. Agardh	Philippines Japan
Enteromorpha sp.	
Monostroma nitidum Wittrock	Japan
	Taiwan, Pr. of China
B. Brown Seaweeds (Phaeophyta)	
Ecklonia sp.	Japan
Eisenia sp.	Japan
Heterochoradaria sp.	Japan
Hizikia sp.	Japan
	Korea, Republic of
Laminaria japonica Areschoug	Japan
L. japonica	China
	Korea, Republic of
Macrocystis sp.	Japan
Nemacystus sp.	Japan
Nereocystis sp.	Japan
Sargassum sp.	Japan
Undaria pinnatifida (Harvey) Sur.	Japan
	China
	Korea, Republic of
U. Peterseniana (Kjellman) Okamura	Japan
U. undariodies (Yendo) Okamura	Japan
C. Bed Seaweeds (Rhodophyta)	
Eucheuma alvarezii Doty	Philippines
E. denticulatum (Burman) Collins et Harvey	Philippines
E. gelatinae (Esper) J. Agardh	China
Gelidium amansii Lamouroux	Japan
Gloiopeltis sp.	Japan

Gracilaria verrucosa (Hudson) Papenfuss	Taiwan, Pr. of China
	Japan
	China
G. gigas Harvey	Taiwan, Pr. of China
G. lichenoides (L.) Harvey	Taiwan, Pr. of China
	Japan
Porphyra angusta Ueda	Taiwan, Pr. of China
P. dentata Kjellman	Taiwan, Pr. of China
P. haitanensis Chang et Zhang Baofu	China
P. kuniedai Kurogi	Korea, Republic of
P. seriata Kjellman	Korea, Republic of
P. suborbiculata Kjellman	Korea, Republic of
P. tenera Kjellman	Japan
P. yezoensis Ueda	Japan
P. quangdongensis Tseng et T.J. Chang	Korea, Republic of China

In Taiwan (PC), Gracilaria is cultured in ponds formerly used for milkfish, with Pingtung County alone accounting for 110 ha of the total 400 ha of Gracilaria ponds in Taiwan (PC) in 1974 and producing 1 000 t of dried Gracilaria seaweed.

In Japan, indoor facilities are used to obtain buds/seedlings for on-growing at sea. The facilities consist of 70-80 cm deep square or rectangular concrete tanks provided with illumination, a temperature control system, and ventilation.

The successful cultivation of seaweeds depends on four important factors:

- Type of seaweeds used

  The seaweeds cultured must be healthy and resistant to disease and breakage. They must be able to grow fast and give high yields during harvest. During processing, they must have high amounts of dry matter from which will be extracted high concentrations of carrageenan of high gel strength and viscosity.

- Ecological conditions of the farm

  The farm must be well-sited and fulfill the bio-ecological requirements of the culture species. In general, the presence of a particular seaweed species in an area is a good indicator of the suitability of that site for culture of the species under consideration.

- Access to sunlight

  Seaweeds being cultivated need abundant sunlight for photosynthesis. Shading by other seaweeds and plants must be prevented by regular inspection and removal of the unwanted plants.

- The seaweed farmer

  The personality and dedication of the seaweed farmer is an important factor since the farmer must visit the farm regularly and carry out routine inspections. Some of the farmer's chores include shaking off silt and other foreign materials from the seaweeds, repairing broken lines, restoring uprooted stakes, and picking up drifting branches of seaweeds.

Trono and Ganzon-Fortes listed the following criteria for selecting good sites for Eucheuma in open waters and Caulerpa and Gracilaria in seawater ponds:

(i) Unpolluted seawater supply.

(ii) Salinity of 30-35 ppt Eucheuma and Caulerpa and 8-25 ppt for Gracilaria.

(iii) Water temperature of 27-30°C.

(iv) Moderate water movement of 20-50 m/min.

(v) Water depth of 0.5-1 m at low tides and not more than 2-3 m at high tides, and

(vi) Firm bottom protected from strong waves for Eucheuma and muddy-loam bottom for Caulerpa ponds.

Seaweeds are grown using different types of planting material (vegetative cuttings, natural seeds, hatchery-reared seeds) and methods of culture (store planting, bottom culture, rope method, rope-concrete method, and pond culture either in monoculture or polyculture with milkfish, shrimp and crabs).

Seaweed Groups/Species	Country Where Cultivated	Type of Planting Material and Methods of Culture
A. Green Seaweeds (Chlorophyta)		
Caulerpa lentillifera J. Agardh	Philippines	Vegetative propagation by cuttings; pond culture
Enteromorpha sp.	Japan	Naturally produced "seeds" grown on hibi nets in open seas
Monostroma nitidumWittrock	Japan Taiwan, Pr. of China	Hatchery-reared or naturally produced "seeds" grown on hibi nets in open seas
B. Brown Seaweeds (Phaeophyta)		
Ecklonia sp.	Japan	Natural seeding on improved substrates
Eisenia sp.	Japan	Natural seeding on improved substrates or introduction of mother plants or seedlings
Heterochoradaria sp.	Japan	No information available
Hizikia sp.	Japan Korea, Rep. of	Introduction of fertile plants on natural or artificial substrates; seeding of naturally produced spores or embryos on rocks
Laminaria japonica Areschoug	Japan	Hatchery produced "seeds"; rope cultivation in open waters using artificial support system; natural recruitment on improved substrates; stone planting or bottom culture using artificially seeded stones

L. japonica	China	Hatchery produced "seeds"; rope cultivation in open waters using artificial support system; scone planting or bottom culture using artificially seeded stones
	Korea, Rep. of	No information (probably same used In Japan)
Macrocystis sp.	Japan	Natural "seeds" on improved substrates; hatchery produced seedlings on twines introduced to artificial substrates
Nemacystus sp.	Japan	No detailed information available
Nereocystis sp.	Japan	No detailed information available
Sargassum sp.	Japan	Introduction of mother plants or seedlings; artificial substrates in open seas
Undaria pinnatifida (Harvey) Sur.	Japan China Korea, Rep. of	Hatchery produced "seeds"; raft or floating rope system in open seas; stone planting using artificially seeded stones; bottom planting in open seas; management of natural stocks by improvement of substrates for natural seeding
U. peterseniana (Kjellman) Okamura	Japan	Same as used for U. pinnatifida
U. undariodies (Yendo) Okamura	Japan	Same as used for U. pinnatifida
C. Red Seaweeds (Rhodphyta)		
Eucheuma, alvarezii Doty	Philippines	Vegetative cuttings using artificial support system on open reefs
E. denticulatum (Burman) Collins et Harvey	Philippines	Same as used for E. alvarezii
E. gelatinae (Esper) J. Agardh	China	Vegetative cuttings tied to pieces of corals and planted on the bottom
Gelidium amansiiLamouroux	Japan	Natural seeding on improved substrates; vegetative cuttings scattered on the bottom and rope-concrete method
Gloiopeltis sp.	Japan	Artificial seeding of substrates using spore suspension or embryos
Gracilaria verrucosa(Hudson) Papenfuss	Taiwan, Pr. of China	Vegetative cuttings; pond monoculture and/or polyculture with milkfish, shrimp and crab
	Japan	Vegetative cuttings Inserted in nets and ropes in protected bays and coves
	China	Vegetative cuttings inserted in bamboo splits; net method; scattering cuttings on the substrate
G. gigas Harvey	Taiwan, Pr. of China	Same as used for G. verrucosa in Taiwan
G. lichenoides (L.) Harvey	Taiwan, Pr. of China	Same as used for G. verrucosa in Taiwan
	Japan	Same as used for G. verrucosa in Japan
Porphyra angusta Ueda	Taiwan, Pr. of China	Hatchery produced seeds; net-raft system in outgrowing areas
P. dentata Kjellman	Taiwan, Pr. of China	Hatchery produced seeds; net-raft system in outgrowing areas
P. haitanensis Chang et Zhang Baofu	China	Hatchery produced seeds on nets using the fixed, semi-floating or floating methods
P. kuniedai Kurogi	Korea, Rep. of	Same as used for P. tenera in Japan

P. seriata Kjellman	Korea, Rep. of	Same as used for P. tenera in Japan
P. suborbiculata Kjellman	Korea, Rep. of	Same as used for P. tenera in Japan
P. tenera Kjellman	Japan	Hatchery produced seeds on bamboo blinds and (recently) on artificially fixed or floating support systems
P. yezoensis Ueda	Japan	Same as used for P. tenera in Japan
	Korea, Rep. of	Same as used for P. tenera in Japan
	China	Same as used for P. haitanensis
P. guangdongensis Tseng et T. J. Chang	China	Same as used for P. haitanensis

Figure: Three methods of Eucheuma culture practised

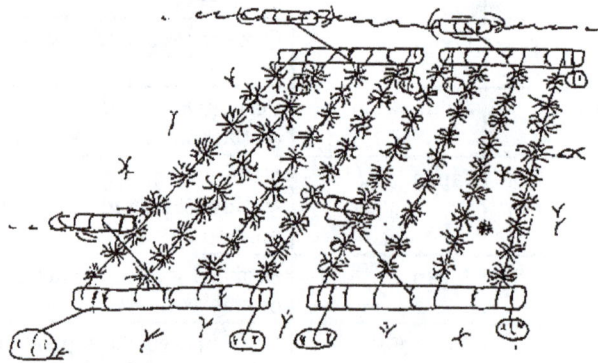

Figure: Three methods of Eucheuma culture practised

In the Philippines, the monoline method of culture is the most popular and successfully used of these methods. The farming activities involved in monoline culture of Eucheuma species based on the Philippine experience are as follows:

(i)   Securing a license from the Bureau of Fisheries and Aquatic Resources (BFAR) prior to farming the area.

(ii) Preparing required materials needed for farm construction.

(iii) Clearing the area of sea grass, seaweeds, large stones and corals, and other foreign materials, followed by measuring it according to the proposed dimensions of the farm.

Wooden stakes are then driven into the bottom with the help of an iron bar and sledgehammer and arranged into 10 m rows at 1 m intervals. An 11 m nylon line is securely tied to one end of each stake about 0.5 m above the ground and then stretched to the corresponding opposite stake and tied securely. If the current is very strong, an additional row of stakes is placed in the middle to provide additional support.

(iv) Obtaining seedlings from the nearest source and transporting them to the farm site within the shortest possible time. During transport, the seedlings are protected from exposure to sun, wind, heat or rain. If the transport of seaweeds will take several hours, the seaweeds are kept damp during the trip and upon arrival at the farm, are immediately submerged in water.

(v) Preparing the seedlings by tying bunches weighing about 50-100 g with soft 25 cm long plastic straw, and then tying these to monolines in the water at 20-25 cm intervals. The plants are allowed to grow to about 1 kg or larger before harvesting.

(vi) Building a farm house if drying of the harvested seaweeds is part of the operations. The farm house is built in or near the farm site so as not to waste time during post-harvest handling. The size of the farm house, which is designed to provide for drying and storage, will depend on the farmer's financial capacity and market commitments.

(vii) Maintaining planted seaweeds by inspecting them regularly while they are growing. Unwanted seaweeds which will compete with the Eucheuma for nutrients and sunlight are removed along with dirt and other foreign materials clinging to the seaweeds. Lost or broken Eucheuma are replaced.

(viii) Harvesting the whole plant and reserving select portions as seedlings for the next crop.

(ix) Sun-drying of the rest of the harvest by spreading these on a drying platform of bamboo slots initially lined with coarse fine-mesh nylon net. The seaweeds are freed of all foreign matter clinging to them.

During hot and sunny weather, it takes about 3-4 days to dry the seaweeds to a moisture content of about 30% or less. The dried materials are then packed in plastic sacks for storage in a dry place or for delivery to the buyer.

The pond culture of Caulerpa involves the following major steps:

• Pond construction

The pond is divided into manageable units measuring about 0.10-0.25 ha. The pond design allows for a flow-through system by providing each unit with its own supply and drainage gates. Water flows uniformly from the main gate to the secondary and exit gates during the draining and flooding process. Peripheral diversion dikes or canals along the landward edge of the pond are also built to divert run-off water from the ponds during the rainy season.

• Planting

To facilitate planting activities, pond water is drained to a depth of about 0.3 m.

Caulerpa seedlings are obtained from the nearest source available and transported to the farm site within the shortest possible time.

The ponds are stocked at a rate of 1 000 kg seedlings/ha or 100 g/m². A handful of seedlings is uniformly buried on one end at approximately 1 m intervals using a string as guide.

After planting, the pond water is gradually raised to a depth of 0.5-0.8 m or just until the plants can be seen from the surface of the water.

The newly planted seaweeds are inspected after a few days. Uprooted seaweeds are replaced and bare areas are replanted.

- Pond management

Water is changed daily or every other day to maintain adequate levels of nutrients. During the initial stages of growth, the seaweeds deplete the water of nutrients at a high rate and frequent water changes are needed to replenish lost nutrients and eliminate the need to fertilize. Water level is, however, carefully maintained to prevent the collapse of the dikes.

Unwanted seaweeds, sea grasses, and animals which will compete with the Caulerpa for nutrients are regularly weeded out.

The dikes and pond gates are inspected regularly to check for leakages, which are repaired immediately. This is vital, especially during the typhoon season.

The application of fertilizer may not be necessary as long as frequent water change is maintained. However, fertilization is resorted to when the stocks appear unhealthy and pale in colour, i.e., from light green to yellowish. When this happens, pond water is changed and fertilizer with a high nitrogen content is applied at the rate of 16 kg/ha by broadcasting or by suspending the fertilizer contained in several layers of plastic sack in strategic areas in the pond. The pond water is not changed in the next two to three days.

- Harvesting

Two months after planting, the Caulerpa forms a uniform carpet on the pond bottom, a good indicator for harvest time.

About 75% of the crop is harvested by uprooting the Caulerpa from the mud and placing it on to a wooden raft.

About 25% of the original crop is left behind, uniformly spaced on the pond bottom to serve as seedstock for the next crop. This may be harvested after two to three weeks.

Harvested seaweeds are washed in clean sea water to remove mud and other dirt. The clean seaweeds are then placed in a basket or clean plastic sheets for further sorting and cleaning before packaging and immediate transport to the market.

Sustainable Aquaculture

Sustainable aquaculture is the cultivation of fish species for commercial purposes by means that have a benign, if not positive, net impact on the environment, contribute to local community development, and generate an economic profit. As a concept, sustainable aquaculture has evolved and grown along with growing evidence that wild fisheries are being overexploited and alarming numbers of fish species are becoming extinct. The negative environmental impact of conventional aquaculture has also motivated those concerned with the oceans, fisheries, and food production to develop a comprehensive definition and set of practitioner's guidelines for sustainable aquaculture. As yet no rigorously defined, universally accepted definition has been agreed upon, nor does an international certification exist.

Aquaculture has been the fastest growing sector of food production worldwide during the past decade. Its growing economic, social, and environmental impacts have led governments, supranational organizations, environmental groups, and industry participants to find more sustainable means of aquaculture development. Consisting of principles and provisions that support this goal, the United Nations Food and Agriculture Organization (FAO) has produced the "FAO Code of Conduct for Responsible Fisheries." Article 9 of the Code addresses development of aquaculture. The essence of the Code emphasizes that fishery resources need to be made use of in a manner that ensures their sustainability over the long term, is in harmony with the natural environment, and does not engage in capture and aquaculture practices that are harmful to ecosystems and communities.

The environmental activism organization Greenpeace, for example, has worked with scientists, researchers, and practitioners to come up with a comprehensive definition of sustainable aquaculture, and one that it is promoting to governments, within the seafood industry, and at international fisheries and environmental conferences. Sustainable aquaculture, according to this definition, strives towards using plant-based feeds farmed using sustainable methods. It avoids fishmeal feeds or feeds based on fish oils from overfished fisheries that result in a net loss in fish protein; nor does it use juveniles caught in the wild.

Sustainable aquaculture also only cultivates open-water species that occur naturally in the location where the aquaculture takes place and then only in bag nets, closed sea pens, or the equivalent; nor does it result in negative impacts to the environment. In addition, sustainable aquaculture has no negative effects on local wildlife or pose threats to local wild populations and does not make use of genetically engineered fish or feed.

There are various other attributes of sustainable aquaculture. It doesn't stock species at densities high enough to risk outbreaks and disease transmission. Nor does it deplete local sources of drinking water, mangrove forests and other natural resources, or threaten human health. It supports local communities economically and socially.

References

- Aquaculture-types-benefits-importance: conserve-energy-future.com, Retrieved 30 June 2018
- Definition-mariculture: lexiconoffood.com, Retrieved 13 April 2018
- Algaculture, fuel-efficiency, biofuels: howstuffworks.com, Retrieved 14 March 2018
- What-is-sustainable-aquaculture: wisegeek.com, Retrieved 09 July 2018

Chapter 2

Flowing and Static Water Aquaculture

In aquaculture, an artificial channel called a raceway is constructed to grow aquatic organisms. It consists of canals or rectangular basins that are made of concrete and has an inlet and an outlet. Fish and other organisms are also cultivated in static water systems such as static freshwater ponds. This is known as static water aquaculture. A detailed analysis of static and flowing water aquaculture has been provided in this chapter, which includes various topics related to static freshwater ponds, water flow in a raceway, raceway systems operation, etc.

Raceway

A raceway in its simplest form is just a flume for carrying water. Raceways for fish culture are tanks which are relatively shallow and rely on a high water flow in proportion to their volume in order to sustain aquatic life. Flow-through fish culture systems pass water through the systems once, provide waste treatment as required, and then discharge the water rather than treating and recirculating it. For successful aquaculture, the inflowing water must be within the temperature tolerance of the species being cultured and should match the optimal temperature for the target species as closely as possible. Oxygen is also provided by the incoming water and is removed by the fish as the water progresses down the raceway. In most raceway systems, dissolved oxygen is replenished by allowing the water to fall into subsequent tanks within the raceway. Dissolved metabolites from animals in the system are carried out in the effluent, while settleable particulate wastes can be captured by settling or less frequently by other means of filtration. Depending on the water chemistry, the depletion of oxygen and the accumulation of ammonia, carbon dioxide, or fine particulates can eventually become limiting to fish production within the system. No natural foods are generated in these systems, and nutritionally complete diets are an essential requirement for successful raceway aquaculture.

Flow-through tilapia farm

Flow-through aquaculture systems require water exchange to maintain suitable water quality for fish production and rely on water flow for the collection and removal of metabolic wastes. Water for flow-through facilities is usually diverted from streams, springs, or artesian wells to flow through the farm by gravity. Water pumped from wells or other sources is more expensive and is seldom used. Water diverted from springs or surface sources for flow-through aquaculture is regulated by various public agencies, depending on the specific water laws of each state. Diversion of surface water is considered a non-consumptive use, although pumping groundwater from a well is considered a consumptive use in some states. The discharge of a high-volume, dilute effluent from flow-through aquaculture facilities greatly limits the treatment options available to producers from both technological and economic perspectives.

Flow-through systems are the most commonly used aquaculture production systems for the culture of rainbow trout Oncorhynchus mykiss and other salmonids in the United States. Other cold-water fish species produced in flow-through systems include brook trout Salvelinus fontinalis and brown trout Salmo trutta. Flow-through systems are used for production of freshwater stages of salmon. Flow-through systems are also used on a limited scale for the production of warmwater fish such as catfish Ictalurus spp. and tilapia Oreochromis spp. Recently, flow-through systems have been used to produce coolwater species such as yellow perch Perca flavescens, hybrid striped bass Morone spp., and several species of sturgeon Acipenser spp.

Flow-through systems include linear earthen and concrete raceways and tanks constructed from other materials. Concrete raceways are the most common. Circular rearing tanks are also used in flow-through systems, most commonly for broodstock production.

The typical raceway production system consists of a tank (rearing unit) or a series of rectangular tanks with water flow along the long axis. In an ideal raceway, water flow will approximate plug flow with uniform water velocity across the tank cross section. However, friction losses at the tank-water and air-water boundary layers will cause water velocities to vary across the width and depth of the raceway. Greatest water velocities are at mid-depth, with slightly reduced velocities at the air-water interface and greatly reduced velocities along the raceway bottom. A defining characteristic of linear-pass raceways is a water quality gradient from the inflow to the outflow of the rearing unit during production, with best water quality at the inflow and deteriorating water quality along the length of the raceway as water flows toward the outlet. Circular rearing units are more thoroughly mixed and have relatively uniform environmental conditions throughout the tank.

Compared to ponds, raceways have several advantages. Per unit of space, raceway production is much higher. Raceways also offer a much greater ability to observe the fish. This can make feeding more efficient, and disease problems are easier to detect and at earlier stages. If disease signs are observed, disease treatments in raceways are easier to apply and require fewer chemicals than a similar number of fish in a pond (due to the higher density in the raceway). Raceways also allow closer monitoring of growth and mortality and better inventory estimates than ponds. Management inputs such as size grading are much more practicable in raceways than they are in ponds, and harvesting is also easier.

The disadvantages of raceways are primarily related to their need for large constant flows of consistent, high-quality water. Since such resources are not common, locating and securing a proper water supply is a major consideration. Also, commercial viability often requires that the water gravity flows through a series of raceways before it is released. This adds a requirement for an elevation of the water source and suitable topography for the gravity flow between raceways.

Another limitation compared to ponds is the release of effluent. While ponds largely process wastes within the culture systems, raceways, with their low retention times, do not. Effluent releases from raceways are a larger consideration than they are for ponds.

Raceway Construction

In-pond Raceways consist of rectangular boxes that can be constructed in various sizes and from several types of materials depending on the intended use. IPRs have been used in research and commercial production at several locations in the South and Midwest since 1992. The smallest IPRs have been used for production of fish fry and were only about 84 cubic feet in volume (6x4x3.5 feet). The largest to date have been used for commercial production of catfish and were approximately 670 cubic feet in total volume (24x8x3.5 feet).

IPRs have been constructed from marine and treated plywood, plastic sheets, and plastic liners. Each of these materials has advantages and disadvantages. Plywood becomes saturated with water and extremely heavy unless coated with non-toxic water-resistant marine paint. Plastic sheets (usually 1 / 4 inch thick) expand and contract with heat, making their shape irregular. Plastic liners (80 mil) cannot be walked in (during harvest or grading) and may collapse due to wave action.

A frame around the outside of the IPR is used for attachment of the plywood or plastic. Both treated lumber and metal frames have been constructed. All IPR materials, including screws and nails, need to be water-resistant and non-toxic. Although treated lumber contains some toxic compounds, these have not been a problem in the IPRs because of the high water exchange rates. However, it may be advisable to coat the wood with non-toxic marine paint.

The IPR is designed to float in any body of water; therefore, a recommended component is a dock or pier for ease of management (e.g., feeding, water testing, etc.). It is possible to anchor the IPR to a stationary pier or to the pond bottom if water levels do not fluctuate. However, if anchored to the pond bottom without a dock, then daily activities must be conducted from a boat. The IPR pier should be constructed of walkways (3 to 4 feet wide) to allow access to all sides of the IPR and provide space for attaching equipment. For ease of management the pier must be constructed so that the IPRs are positioned close to the walkways. Security should also be considered in construction. Theft and vandalism can be a problem in any type of high density fish culture system.

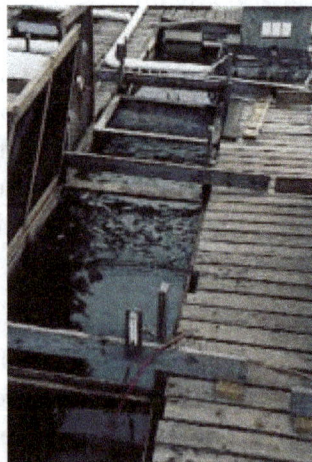

Figure: IPR with dock–note that blower housing and air-manifold are at the forward end of the raceway

One of the most common IPRs has been built of treated plywood, framed with treated 2x4 lumber or steel, and coated with epoxy paint. Sizes have been either 16 or 24 feet long, 4 feet wide, by 4 feet deep (only 3 feet underwater).

Figure shows the basic design of this raceway. The air-lift pump is attached to the front or "intake" end of the IPR, and the waste col- lection system (if needed) is attached to the "discharge" end. The intake end wall of the race- way is constructed so its upper edge is approximately 9 inches below the sides of the raceway. This space allows the air-lift pump to be adjusted for flow control. The rear discharge wall of the raceways is constructed so that its lower edge ends about 4 inches above the raceway bottom. This allows discharged wastes to be drawn off the bottom of the race- way for removal.

An "eddy board," usually 2x6 or 2x8, is placed across the width of the raceway about 4 to 6 feet from the water discharge of the air-lifts. This board should be attached with about 1 inch extending above the water surface when the pumps are running. The eddy zone behind this board is the feeding area of the raceway. Feed dropped in this area is held against the board, keeping it from being washed out of the raceway. Cage-type mesh material (usually 1 / 2-inch mesh) is used to keep cultured fish inside the IPR and exclude wild fish from entering, without restricting water flow.

Mesh is placed in front of the air- lifts and at the discharge end of the raceway. The mesh in front of the air-lifts should be in an "L"- shape, forming a trough across the raceway about 6 to 8 inches in front and 4 to 6 inches below the air-lift's water discharge. This trough traps debris and wild fish that enter through the air-lifts, without restricting water flow. A second mesh screen is placed about 1 foot from the rear of the raceway and extends completely across the width and height of the raceway. The rear screen keeps the cultured fish from leaving the raceway and wild fish from entering through the water discharge opening.

Figure: IPR (out of the water) showing position of air-lift system and eddy board

Hinged lids or doors should cover the top of the IPR to discourage predators and stop fish from escaping by jumping out. Usually several small lids are preferable to one or two large ones because of weight and the need to access only certain sections of the race- way at a time. Mesh material should be used in the section over the feeding area so feed can be dropped into the race- way without opening the lid. The remaining lids can be constructed of a solid material or can be covered with a material such as shade cloth to reduce light and its associated stress on the fish. A 16x4x4 foot IPR has an effective culture volume of 210 cubic feet (15x4x3.5 feet) or 1,571 gallons. A 24-foot- long IPR would have an effective culture volume of 322 cubic feet (23x4x3.5 feet) or

2,409 gallons. An IPR with dimensions of 24x8x4 has an effective culture volume of 644 cubic feet (23x8x3.5 feet) or 4,817 gallons.

Figure: IPR (out of the water) built from treated plywood showing external frame and attachment of tube settler.

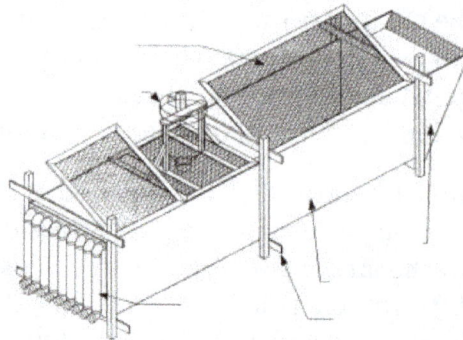

Figure: Drawing of IPR showing attachment of airlifts, tube settler, lids, and demand feeder

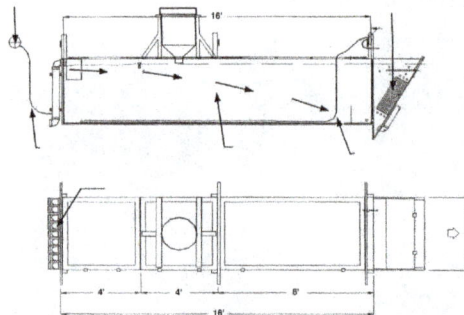

Figure: Drawing of IPR in cross section and top view showing attachment of air-lifts, tube settler, demand feeder, and emergency oxygen tubing system

IPRs have also been constructed using plastic liners. There are several synthetic compositions (i.e., chemically different) of plastic liners. These are commercially avail- able in 19 to 80 mil thickness, with 40 mil being adequate for use in most IPR situations. Liners are ultraviolet light-resistant and have a lifetime of at least 10 years. Liner manufacturers can fashion liners in many shapes and sizes, so it is possible to have a liner custom-made for a specific IPR design. The flexible nature of a liner allows the raceway to be moved to the pond bank or pier and collapsed for easy harvesting of the fish.

A disadvantage of plastic liner construction is that the walls can collapse inward from wave action, reducing the raceway volume unless a frame is used to maintain its shape. Also, the attachment of solid waste collection systems and air-lifts is more difficult since it is hard to glue materials to the liner. Cost of plastic liners is also a consideration. Depending on thick- ness, the type of liner, and custom shaping, they can range in price from $0.50 to several dollars per square foot. This cost is only for the liner and does not include frame, blower, air-lifts, etc.

A small IPR (8x3x3 feet) of 23 mil plastic liner has been tested for use as a fry rearing unit. For this purpose an IPR offers advantages as mentioned before and also pro- vides a steady supply of plank- tonic food organisms essential for good growth and survival of fry. A saran mesh sock of 250 microns was placed over the air-lift dis- charge or outflow to prevent any predaceous insects or fish from entering the raceway. The saran sock will not screen out plankton in the water. Prob- lems were encountered with fouling of the solids from the bottom of the race- way. Results of this study on fry production were promising, how ever.

Air-lift Pumps

Air-lifts provide a simple and efficient means of moving large volumes of water. Rising air bub- bles inside an air-lift's tube act like a piston pushing water above it. However, this is efficient only if the water is lifted a small height above the surface. In fact, most air-lifts will not lift water over 3 or 4 inches above the water's surface. Air-lifts work most efficiently when they are releasing water at or very near the surface.

Air-lifts have the added benefit of aerating incoming water when dissolved oxygen (DO) concen- trations are much below saturation. In research trials, when pond DO fell below 2 mg / L the DO in the IPRs has been maintained at 3 mg / L even with high biomass. Because of the mixing action of water and air in the air-lifts, super saturation is virtually eliminated in the water entering the IPR.

Air-lift pumps consist of a battery of single air-lifts. Individual air- lifts are constructed from a 36- inch long section of 3- or 4-inch PVC pipe. A 4-foot-wide raceway has room for the attachment of 9 3-inch diameter air-lifts. A PVC 900 elbow or "L" is glued to the top of each air-lift. Each air-lift is designed so that air from the blower enters the pipe at approximately 32 inches below the center of the PVC "L". Regenerative blowers are most efficient at powering air-lifts if the air is injected between 30 and 34 inches below the surface of the water.

Figure: Air-lifts attached to a sheet of plywood at the front of raceway, showing garden hose attachment at a water depth of 32 inches

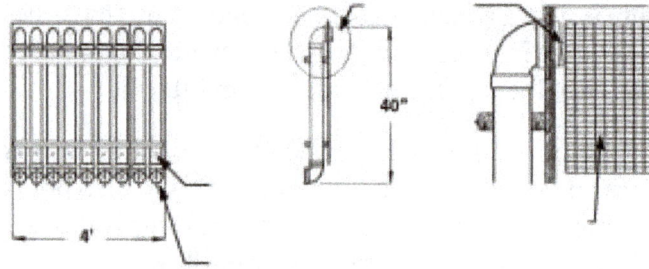

Figure: Drawing of air-lift pump design showing attachment of individual air-lifts to plywood or plastic sheet

900 elbow or "L" is glued to the top of each air-lift. Each air-lift is designed so that air from the blower enters the pipe at approximately 32 inches below the center of the PVC "L". Regenerative blowers are most efficient at powering air-lifts if the air is injected between 30 and 34 inches below the surface of the water.

Optimally the air-lifts are sub- merged to the halfway point of the "L" or to the top of the "L". Each air-lift is attached to a ply- wood or plastic panel. A circular cut-out is made so that each "L" protrudes through the panel and into the raceway area. Silicon sealer can be used around the cut- outs to seal the "L's" to the panel.

This keeps water from escaping the raceway around the "L's". Air-lifts are attached to the panel with screws or by pressure straps. Bolts or screws should not extend into the pipe more than 1 / 2 inch, as debris can become caught on this obstruction and reduce water flow through the air-lift. Each air- lift in an IPR system must be built identically to all others and attached to the same air-manifold and blower in order to work properly.

The panel to which the air-lifts are attached fits into tracks along each side at the front of the raceway. These tracks allow the air-lift pump to be raised or lowered to adjust the height and therefore the water flow through the air-lifts.

The intake of the air lift should be approximately 36 inches under- water. The intake can be moved upward or downward to utilize different water temperatures or conditions. For example, if warmer and more oxygen-rich surface water is desired, the intake could be turned upward (starting at the bottom of the 36- inch vertical section) using elbows and pipe to place the intake closer to the water's surface. A longer vertical extension could be used if cooler water was desirable. This would depend upon the quality of the deeper water.

Air is supplied to the air-lift pumps by a regenerative blower. Regenerative blowers are high-volume, low-pressure units. The blower is attached to an air-manifold that holds a large volume of air under constant pressure. With out the proper volume in the air- manifold the air-lifts will not function effectively, and the regenerative blower will be damaged due to overheating. Typically a 1-horsepower blower requires a minimum of 20 feet of 4-inch PVC or 12 feet of 6-inch PVC air-manifold (approximately 2,500 cubic inches). One-half-inch PVC tubing connectors are tapped into the air- manifold and into the air-lifts (at 32 inches as described previously). A section of garden hose (5 / 8 inch), polypropylene, or plastic tubing (1 / 2-inch ID) can be used as air-line between the air- manifold and the individual air- lifts. The air-line attaches over the tubing connectors from the air- manifold to each air-lift.

The key to making all the air- lifts work properly is that they all must be constructed exactly alike, and each requires a constriction orifice at the attachment of the air-line to the air-manifold. The constriction orifice should have a 3 / 16- to 1 / 4-inch hole in its center. This orifice can be made from PVC or Plexiglas sheeting (1 / 8 to 1 / 4 inch thick) and hot-glued to the PVC tubing connector. If constructed in this fashion, a 1-horsepower blower can efficiently power 27 individual air-lifts or enough for 3 separate 4-foot-wide raceways with 9 air-lifts each.

Figure: PVC tubing connectors with restriction orifice

Water flow through the IPR(s) with this air-lift pump design can be regulated by raising or lowering the air-lift pump, or by stop- ping the air flow to individual air- lifts. With all 9 air-lifts functioning properly the flow rate aver- ages about 450 gallons per minute. At this flow rate a 16x4x3.5-foot raceway completely flushes in less than 4 minutes. At this flushing rate the carrying capacity of the 9 air-lift IPR appears to be approximately 3,000 pounds with warmwater species (e.g., catfish), a stocking rate of 13.4 pounds per cubic foot.

Air-lift pumps have also been constructed in a box or square design. In this type of pump a box is made from plywood or plastic panels 3 inches wide with vertical partitions every 3 inches, resulting in a unit with each individual air-lift a 3-inch square tube, 3 feet long. Air injection, water discharge, screening, and vertical slide adjustments are similar to those described for the PVC air-lifts above. This design allows as many as 13 air-lifts in a 4-foot- wide area.

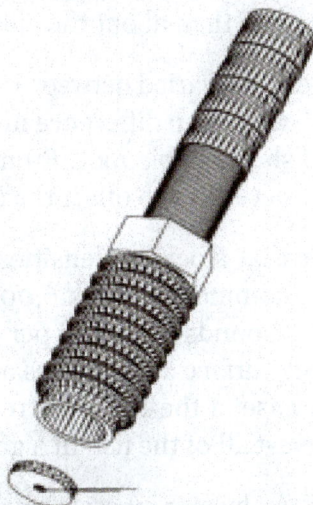

Figure: Drawing of PVC tubing connector with restriction orifice

Emergency Systems

The IPR needs emergency back-up systems in case of electrical disruptions or mechanical failures. A backup blower is recommended in case of blower failure. In addition, the two blowers can be equipped with a pressure sensor that will turn on the backup blower in the event of a failure. Sensors can be purchased that will sense not only power failures but air- pressure loss (in the case of a cracked air-manifold). These sensors can be attached to phone dialers which will call managers and alert them to problems and can automatically trigger emergency generators or oxygen supply systems.

A simple oxygen supply system can be constructed using cylinders of bottled oxygen connected to a normally-closed electric solenoid valve that opens if electrical power is interrupted. High pressure tubing leads from the cylinders to each raceway and is delivered through milli-pore tubing in the bottom of each raceway, similar to a hauling tank system. Flow regulators control the volume of oxygen delivered and must be adjusted depending on the biomass of fish in the raceways.

Typically a single cylinder of oxygen will maintain a raceway for several hours. This system is also used to maintain adequate oxy- gen supplies during therapeutic bath treatments for disease.

Species and Stocking Rates

To date, species that have been successfully cultured in IPRs include: channel, blue, and hybrid catfish; trout, striped bass and its hybrids, yellow perch, bluegill, and tilapia. Probably any species that tolerates flowing water can be cultured in an IPR.

Channel catfish and Nile tilapia have been successfully polycultured in the IPR. In one experiment tilapia were mixed in the IPR at a 1:10 ratio with catfish. In other experiments tilapia were isolated in a separate section of the IPR behind the catfish and were not fed, under the assumption that they would eat any uneaten catfish feed, catfish wastes, and plankton. The tilapia grew well in both these experiments.

Blue catfish and channel X blue hybrid catfish did not perform as well in the IPR as channel catfish in experiments at Auburn University. However, producers in more northern climates have reported success in culturing these in raceways. These observations may indicate more about the temperature preference of the blue catfish than about the culture system.

Stocking rates for most of these species have varied between 9 and 15 fish per cubic foot of effective culture volume. At least in the case of catfish, no difference in growth or food conversion has been found between stockings at 9 or 14 fish per cubic foot. From an economic standpoint, the high stocking rates of the IPR are probably necessary to offset the cost of construction and operation.

Finally, it is important to remember that stocking densities must be balanced with pond size. In open-pond catfish production it is common to stock 6,000 or more fish per surface acre but expect to harvest only 3,500 to 4,000 pounds of catfish per year. In cages, catfish are normally stocked at only 1,500 to 2,000 fish per surface acre (unless aeration is supplied), and all the fish are harvested in a given year. In the case of the IPR, it is recommended to stock no more than 6,000 fish per acre and expect to harvest all of the fish in a given year.

As a note of interest, several species of freshwater mussels have also been cultured behind catfish in the IPR in an attempt to reduce effluent wastes. The mussels were somewhat effective at reducing solid wastes in the effluent, and some species of mussels showed significant growth under these conditions. This research may have implications for the culture of freshwater mussels (the shells of these species are used as nuclei for cultured pearls) or in the culture of other shellfish species in brackish or marine environments.

Feeding

Feeding rates (percent body weight per day) and times depend more on species cultured than on the culture system. Floating feed is recommended for the IPR, because the manager can see fish eat and determine if any feed is being wasted or uneaten. The IPR does allow the use of sinking feeds, including medicated feed if necessary.

Traditional raceway culture has often utilized demand or automatic feeders. Research on catfish in the IPR has shown that demand feeders work well. In fact, with catfish and tilapia there were no differences in growth or feed conversion using demand feeders as compared with twice- a-day hand feeding.

Fish cultured in raceways have better feed conversions than fish grown in open ponds with the possible exception of tilapia. This is also true of the IPR. In 5 years of research on catfish and tilapia, the average feed conversion ratio (FCR) was 1.45:1 (pounds feed fed to pounds of fish produced).

Finally, because of the high density and lack of any natural foods, raceway culture depends on high quality complete diets. In IPR research on catfish and tilapia at Auburn University, a 36 percent protein commercially available diet was fed in most experiments, rather than the 32 percent protein diet that is commonly used in pond culture. Most cage producers also use a 36 percent protein complete diet.

Disease Treatments

Disease treatments in raceways are usually drip treatments. The therapeutant is dripped into the incoming water, and a specific concentration is maintained for a certain period of time, usually 1 hour. Problems with this method are that the concentration is difficult to maintain, a large amount of therapeutant is used, and therapeutant is released into the environment with the discharge.

In the IPR, the emergency oxygen system can be used to conduct therapeutic bath treatments. In this case the air blower is turned off and the emergency oxygen supply system is used. With no water flow the raceway is treated as a tank of known volume. The therapeutant is mixed into the raceway at the prescribed concentration and maintained for the recommended time period. DO concentrations should be checked and the oxygen supply regulated during the treatment. After treatment the air blower is turned on, and the therapeutant is flushed out of the raceway within a few minutes. Obvious advantages of this system are that less therapeutant is used, a more precise con- centration is achieved, and if problems occur the treatment can be terminated quickly.

Waste Reduction

One of the anticipated benefits of the IPR was to capture or reduce wastes from the system. By doing this the IPR system would be more "environmentally friendly" and / or could produce more fish per acre, particularly when com- pared to cages in watershed ponds. However, it should be noted that fish wastes are mostly soluble, and solids are almost neutrally buoyant and therefore difficult to settle. If the pond utilized is large and the stocking density per acre low, it may not be necessary to practice waste reduction at all, since the pond should be able to absorb and decompose the waste effluent through natural cycles.

Several different low cost and low maintenance methods of trapping or reducing wastes from the IPR have been researched. These have included settling basins, tube settlers, sand and synthetic mesh filters, plant and gravel biofilters, artificial wetlands, and filter-feeding species in polyculture. The best methods appear to be poly- culture with filter feeding species, and tube settlers (for the solids) coupled with some type of plant biofilter or artificial wetland outside of the raceway. Actual costs of these waste reduction systems and their total impact on the pond environment have not been adequately evaluated.

Figure:. Tube settler constructed of 3/4-inch schedule 20 PVC

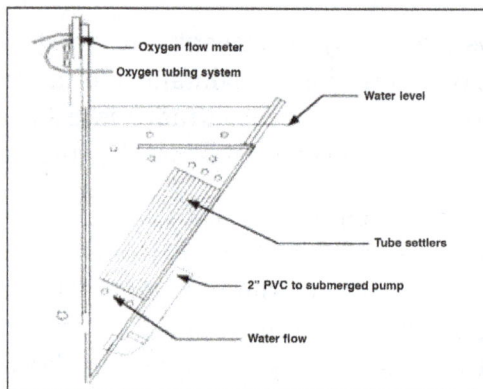

Figure: Tube settler drawing showing construction with schedule 20 PVC and attachment of oxygen flow meter

Problems

All culture systems have advantages and disadvantages. Like other high density raceway systems the IPR has problems related to disease, reaction time, and predators.

Diseases, particularly bacterial diseases, are common in all high density systems, especially raceways, cages, and recirculating systems. Bacterial diseases, particularly Enteric Septicemia of Catfish (ESC) and Columnaris, have been problematic with the IPR catfish research at Auburn University.

Survival of catfish in IPR research has ranged from 65 to 98 percent, which is similar to cage research in the same pond. Tilapia survival has averaged around 97 percent; most of these losses have been due to escapement. Commercially operated IPRs have reported bet- ter over- all survival.

Reaction time is another problem with the IPR as with other high density production systems.

Backup systems, either generators or pure oxygen systems, are absolutely essential as power disruptions are inevitable. Since generators eventually run out of fuel and oxygen cylinders become depleted, electrical and / or pressure sensors with phone dialers are prudent components of these systems.

Predators, particularly birds, raccoons, and otters, are attracted to IPRs. The lids and mesh barriers around the inflows and outflows must be properly constructed and routinely maintained to exclude these persistent predators.

Economics

Cost of constructing an IPR system can vary greatly depending on the size and the materials used. The IPR has also been utilized as an effective fish holding system. Several small processors of catfish have built and used IPRs to hold fish for later processing or live sales. They report that the fish adapt immediately to the IPR without any associated trauma, as usually occurs if large catfish are placed in cages. Feeding can also be started to continue weight gain or maintain the weight of the fish if they are to be held for a long time. The same processors have reported that catfish have been purged of off-flavor in the IPR within a few days to a week, as long as the pond in which the IPR was located did not have an off- flavor episode during the purging period.

Water Flow in a Raceway

Among all systems of fish culture, the flow-through system depends to the highest degree upon an abundant and continuous water supply.

In a conventional flow-through system the oxygen requirement of the fish is supplied by the inflow water. The water flow rate that is needed for proper oxygen supply of the fish usually is larger than is required for flushing the metabolic wastes. Thus, in a conventional flow-through system the water flow rate should be calculated on the basis of the oxygen requirement of the fish. Recently, intensive flow-through systems have been designed in order to increase the stocking density or to decrease the water flow. In these systems the oxygen requirement of the fish is met by oxygenation of the inflow water. When the water flow rate of an intensive flow-through system is calculated, the flow rate that is needed for the flushing of the metabolic wastes becomes the critical factor.

Conventional Flow-through Systems

The specific flow rate (q) is one of the basic parameters of flow-through systems that can be expressed:

$$q = \frac{Q}{W} \, (m^3 h^{-1} kg^{-1})$$

In this formula and all subsequent formulas, the notation h-1 denotes per hour, kg-1 per kg, etc.

where

Q = actual water flow (m h)

W = actual mass of fish in the tank (kg)

In the conventional flow-through system the oxygen requirement of the fish stock is ensured by the inflow water as follows:

$$Q \cdot (Cs - C) = W \cdot r$$

$$\frac{Q}{W} = \frac{r}{C_s - C}$$

where

Cs = dissolved oxygen concentration at saturation level (when the oxygen is dissolved from the atmosphere) (gm-3)

C = allowable minimum dissolved oxygen concentration (gm-3)

r = specific oxygen consumption of the fish (gh-1kg-1)

Assuming the following basic data:

- Temperature of inflow water is 15°C

- Inflow water is saturated with oxygen (C = 10 gm-3)

- Specific oxygen consumption of the fish is 0.4 gh-1kg-1.

The specific flow rate can be calculated as follows:

$$q_{(15)} = \frac{Q}{W} = \frac{r}{C_s - C} = \frac{0.4}{10 - 5} = 0.08 m^3 h^{-1} kg^{-1} = 1.92 m^3 \, day^{-1} kg^{-1}$$

There is another important parameter that can be used for the design of flow-through systems, the specific volume (V) of water necessary for a 1 kg weight increase of fish. This can be expressed:

$$V = \frac{q}{G} = (kg \, m^{-3})$$

where

G = specific growth rate (kg kg-1 day-1)

G can be expressed by the known equation as follows:

$$G = \frac{1}{W} \cdot \frac{dW}{dt} = R_p \cdot PER - h (kgkg^{-1}day^{-1})$$

where

Rp = Food protein consumption of 1 kg of fish per day (kg kg day)

PER = Protein Efficiency Ratio (kg fish flesh/kg food protein)

h = daily fish mortality rate (kg kg-1 day-1)

assuming,

Rp =0.01 kg kg-1day-1

PER = 2kg kg-1

h = 0.01kg kg-1day-1

Then G = RpPER - h = 0.01 . 2 - 0.01 = 0.01 kg kg-1day-1

The specific volume of water (V) can now be calculated as follows:

$$V = \frac{g_{(15)}}{G} = \frac{1.92}{0.01} = 192\,m^3kg^{-1}$$

If the temperature of the inflow water is not 15°C but 20°C (Cs = 9 gm), the specific values are as follows:

q (20) = 2.4 m3day-1 kg-1

v (20) = 240 m3kg-1

Intensive Flow-through Systems

The value of specific water consumption can be decreased by supplying pure oxygen to-the inflow water. When the pressure is 100 kPa and the water temperature is 15°C, 48 gm of pure oxygen can be dissolved in clean water. $(C_s^x(15)=48\,gm^{-3}$. At a water temperature of 20°C, the $C_s^x(20)=43.6\,gm^{-3}$.

If the minimum allowable oxygen concentration is 5 gm-3 ,1 m3 of water contains 38.6 g dissolved oxygen available for fish at 20°C which is almost tenfold that contained by water with oxygen dissolved from the atmosphere:

$$\frac{C_s^x - C}{C_s - C} = \frac{43.6 - 5}{9 - 5} = 9.65$$

The specific flow rate (q) at two different temperatures is the following:

q (15) = 0.223 m3 day-1kg-1

q (20) = 0.248 m3 day-1kg-1

The specific volume of water (V) is the following:

v (15) = 22.3 m3kg-1

v (20) = 24.8 m3kg-1

These data, however, should always be checked as to whether toxic metabolites accumulate at these flow rates.

Our assumption is that fish can incorporate some 30 percent of the feed protein (PPV = 0.3), the rest, 85 percent of which is ammonia, is excreted. Since 16 percent of the protein is nitrogen, fish excrete 95.2 g ammonia for each kg of feed protein consumed.

The calculation of the amount of ammonia-nitrogen excreted after feeding a unit of food protein (a) is as follows:

a = (1 - PPV) 0.85 . 0.16 (kg kg-1)

where

PPV = Productive Protein Value (kg fish protein/kg feed protein)

a = (1 - 0.3) 0.85 . 0.16 = 0.0952

Ammonia, depending on the pH and temperature of water, is present in the water in two forms:

$$NH_3 + H_2O \rightleftharpoons NH_4^+ + OH^-$$

as ammonium ion (NH_4^+) and as 'free' or 'un-ionized' ammonia (NH_3) that is toxic to the fish. In our calculation the allowable maximum value of un-ionized ammonia concentration is 0.045 gm-3.

Calculating the tolerable total ammonia-nitrogen concentration (C) at different pH and temperature values yields the following:

CN/20°C, pH 7.01/ = 10.71 gm-3

CN/l5°C, pH 7.5/ = 4.75 gm-3

CN/20°C, pH 7.5/ = 3.43 gm-3

CN/25°C, pH 7.5/ = 2.40 gm-3

CN = ammonia-nitrogen concentration (gm-3)

$(NH_3 - N) + (NH_4^+ - N)$

For the calculation of CN, figure and table below should be used.

In intensive flow-through systems an adequate water flow is needed in order to flush the metabolic wastes, first of all the ammonia.

The amount of ammonia + ammonium ion that is excreted by a certain mass of fish (W) during a day can be expressed as follows:

1000 . W . a . Rp (g day-1)

where

W = mass of fish in a tank (kg)

a = amount of ammonia-nitrogen excreted after feeding a unit of food protein (kg kg-1)

Rp = fish feed protein consumed by a unit mass of fish in one day (kg kg-lday-1)

The amount of ammonia-nitrogen that can be flushed by the water flow (Q) in a day, with a given maximum tolerable ammonia-nitrogen concentration (CN) can be expressed as follows:

24 . Q . CN (g day-1)

where

Q = water flow (m3 hour-1)

CN = ammonia-nitrogen concentration (gm-3)

thus,

$$1\,000 . W . a . Rp = 24 . Q . CN$$

Dividing both sides of the equation by W and using the formula:

q = Q . W , we get an equation as follows,

$$q = \frac{1000}{24} . \frac{aR_p}{C_N} \ (\mathrm{m^3 hour^{-1} kg^{-1}})$$

In our calculation:

$$q = \frac{1000}{24} . aR_p = \frac{1000}{24} . 0.0952 . 0.01 = 0.396$$

Thus, the specific flow rate (g) at different pH and temperature can be calculated as follows:

$$q\ (15°C, pH\ 7.5) = \frac{0.0396}{4.75} = 0.200 \ \mathrm{m3day\text{-}1kg\text{-}1}$$

$$q\ (20°C, pH\ 7.5) = \frac{0.0396}{3.43} = 0.278 \ \mathrm{m3day\text{-}1kg\text{-}1}$$

The specific volume of water can be calculated as follows:

$$V (20°C, pH\ 7.5) = \frac{q}{G} = \frac{0.0278}{0.01} = 27.8\ m3\ kg\text{-}1$$

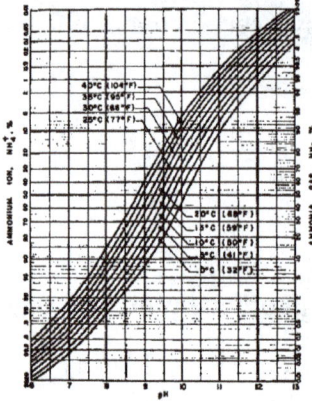

Figure: Ammonia equilibrium pH and temperature

Temp. (°C)	6.0	6.5	7.0	7.5	8.0	8.5	9.0	9.5	10.0
0	.00827	.0261	.0826	.261	.820	2.55	7.64	20.7	45.3
1	.00899	.0284	.0898	.284	.891	2.77	8.25	22.1	47.3
2	.00977	.0309	.0977	.308	.968	3.00	8.90	23.6	49.4
3	.0106	.0336	.106	.335	1.05	3.25	9.60	25.1	51.5
4	.0115	.0364	.115	.363	1.14	3.52	10.3	26.7	53.5
5	.0125	.0395	.125	.394	1.23	3.80	11.1	28.3	55.6
6	.0136	.0429	.135	.427	1.34	4.11	11.9	30.0	57.6
7	.0147	.0464	.147	.462	1.45	4.44	12.8	31.7	59.5
8	.0159	.0503	.159	.501	1.57	4.79	13.7	33.5	61.4
9	.0172	.0544	.172	.542	1.69	5.16	14.7	35.3	63.3
10	.0186	.0589	.186	.586	1.83	5.56	15.7	37.1	65.1
11	.0201	.0637	.201	.633	1.97	5.99	16.8	38.9	66.8
12	.0218	.0688	.217	.684	2.13	6.44	17.9	40.8	68.5
13	.0235	.0743	.235	.738	2.30	6.92	19.0	42.6	70.2
14	.0254	.0802	.253	.796	2.48	7.43	20.2	44.5	71.7
15	.0274	.0865	.273	.859	2.67	7.97	21.5	46.4	73.3
16	.0295	.0933	.294	.925	2.87	8.54	22.8	49.3	74.7
17	.0318	.101	.317	.996	3.08	9.14	24.1	50.2	76.1
18	.0348	.109	.342	1.07	3.31	9.78	25.5	52.0	77.4
19	.0369	.117	.368	1.15	3.56	10.5	27.0	53.9	78.7
20	.0396	.125	.396	1.24	3.82	11.2	28.4	55.7	79.9
21	.0427	.135	.425	1.33	4.10	11.9	29.9	57.5	81.0
22	.0459	.145	.457	1.43	4.39	12.7	31.5	59.2	82.1
23	.0495	.156	.491	1.54	4.70	13.5	33.0	60.9	83.2
24	.0530	.167	.527	1.65	5.03	14.4	34.6	62.6	84.1
25	.0569	.180	.566	1.77	5.38	15.3	36.3	64.3	85.1
26	.0610	.193	.607	1.89	5.75	16.2	37.9	65.9	85.9
27	.0654	.207	.651	2.03	6.15	17.2	39.6	67.4	86.8
28	.0701	.221	.697	2.17	6.56	18.2	41.2	68.9	87.5
29	.0752	.237	.747	2.32	7.00	19.2	42.9	70.4	88.3
30	.0805	.254	.799	2.48	7.46	20.3	44.6	71.8	89.0

Figure: Percent NH in Aqueous Ammonia Solutions for 0-30°C and pH 6-10

The values of the specific flow rate and the specific volume of water at different pH and water temperatures are shown in table. It can be seen in the table that at 15°C the specific flow rate needed for the proper oxygen supply is higher than that needed for the flushing of the metabolic wastes. At higher temperatures the ammonia removal by flushing becomes the decisive factor. It also turns out from the table that the water requirement of the system is two times higher at 20°C than at 15°C.

Raceway System Operations

Feed must be stored onsite, and guidelines for feed storage from Aquaculture Best Management Practice (BMP) No. 7 should be implemented. Therapeutic agents often are used in trout culture, and practices in BMP No. 11 should be implemented regarding storage, use, and disposal of these products. Moreover, BMP Nos. 14 and 15 should be followed regarding general operations, worker safety, and emergency response.

In addition to the practices mentioned above, certain practices specific to flow-through systems should be applied as follows:

- Management plans should be prepared by and practices implemented with the assistance of a professional engineer (PE) licensed in the State of Alabama or other qualified credentialed professional (QCP). Periodic inspections of the operation also should be conducted by a PE or QCP.

- Feed should be offered several times per day in quantities that the fish will completely consume. The maximum daily feed input should be based on the relationship that 1 lb feed requires 0.2 lb oxygen.

- Fish should be excluded from the last 6 to 8 ft of each raceway unit by a screen. This will allow an area for sedimentation of uneaten feed and feces.

- Sediment should be removed from the ends of raceways at 1- or 2-day intervals by suction or via a center drain for treatment in a sedimentation basin. Any sediment that is removed should be disposed in a responsible manner according to NRCS technical standards and guidelines.

- Dead fish should be removed daily for disposal in a responsible manner according to NRCS technical standards and guidelines.

Implementation Notes

Experience with trout and other species in flow-through systems indicates that for each pound of feed applied, 0.2 lb of dissolved oxygen will be needed. Dissolved oxygen concentration should not decline below 5 ppm in trout raceways. Thus, if the incoming water contains 9 ppm dissolved oxygen, only 4 ppm dissolved oxygen (9 ppm in inflow – 5 ppm in outflow) is available. The weight of feed that may be applied daily can be calculated with the equations:

(1) $Inflow(m^3/min) = Inflow(gpm) \times 0.003785.$

(2) $Inflow(m^3/min) = Inflow(ftH^3/sec) \times 1.7.$

(3) $Weightfeed(lb/day) = Inflow(m^3/min) \times (g/m^3\ DO\ in\ inflow - 5g/m^3) \times 1,$

$440\,min/day \times 0.001\,g/kg \times 5kg\ feed/kg\ oxygen \times 2.205\ lb/kg.$

For example, suppose a flow-through system has an inflow of 2,000 gpm containing 9 ppm dissolved oxygen. The maximum daily feed input is calculated as follows:

$Inflow = 2,000\,gpm \times 0.003785\,m^3/min = 7.57\,m^3/min$.

$Weight\,feed = 7.57\,m^3/min \times (9g/m^3 - 5g/m^3) \times 1,440\,min/$

$day \times 0.001g/kg \times kg\,feed/kg\,oxygen \times 2.205\,lb/kg = 481\ lb/day$

The calculations are much easier with metric units than with English units. Please note that,

$1ppm = 1mg/L = 1g/m^3$

The available dissolved oxygen can be increased by providing gravity aeration that allows water to fall between raceway units or by applying mechanical aeration or oxygenation. For warmwater fish culture in flow-through systems, the dissolved oxygen concentration can be allowed to decline to 4 ppm without stressing fish.

The separation of particulate material from water is achieved by utilization of basic differences between the physical properties of the particulates and the encompassing water. For example, differences in particle size, density, biological responses, electri- cal attributes, magnetic properties, or chemical characteristics have all been exploited in various solids removal or separation operations. An ideal solid-liquid separation system would result in a stream of liquid going one way and dry solids going another. Practical systems, however, typically provide varying degrees of solids separation.

Reducing Inputs

In addition to water treatment methods to separate particles from the water, a nutritional approach is another strategy for decreasing the solids discharged in an aquaculture effluent. Reducing the feed inputs into the system or improving the conversion of inputs into fish biomass is a fundamental way of decreasing the solids discharged from an aquacultural operation. Investigators have observed that fish in controlled laboratory situations typically exhibit maximum growth at lower feed conversions than in farm situations. The difference in feed conversion is largely a result of wasteful feeding practices at the farm sites. Studying land-based salmon farming tanks in Norway, Bergheim determined that feed losses represented about 50 percent of the total pollution loading. Seymour and Bergheim looked at net pen culture in Norway and estimated that about one-third of the supplied feed was wasted, and that wasted (uneaten) feed contributed about twice the solids loading as fecal production. Chen undertook a field survey of water quality parameters during a growing season and concluded that excessive stocking density, poor siting and consistent overfeeding contributed to serious organic pollution and eutrophication in the area around cage culture sites of rock cod and other species in the Dongshan region of Fujian province, China. Using both surface and underwater observation to study feeding practices in the net pen culture of salmonids, Ang and Petrell noted that feed loss through the water column and high energy swimming patterns could be avoided by altering feeding methods. The authors recommended a more continuous feeding regime to promote a more orderly and less energy intensive foraging behavior in the fish. Another recommendation was to match feeding rate to foraging success rate rather than supplying a fixed ration. Ang and Petrell (1998) also noted, based on their underwater observations, that feed pellet wastage increased when visibility dropped below 3 m.

A variety of nutritional and feeding approaches can contribute to minimizing the solids loading in the effluent. Improved feed handling and distribution methods reduce the introduction of partially disintegrated pellets and pellet dust into aquaculture systems. Increased stability of feed pellets, through processing methods or added binders, lengthens the time that an intact pellet can be taken by the fish or captured by a solids-removal device. Feeding strategies optimized for a given species and growing condition help to maximize uptake by the fish and maintain consistent water quality conditions. Feed formulations matched to the nutrient requirements and life stage of the fish work to minimize the excretion of excess nutrients.

"Input-based" or supply-side strategies to reduce effluent loading is an important area of investigation that complements the variety of other approaches to minimizing the environmental impact of aquaculture. Coupling the outputs from an aquacultural operation to systems that utilize those outputs in a benign manner is one such strategy.

Coupled-system Approaches

One person's trash is another's treasure. Utilizing the "waste" products from one process as the inputs to another is a fundamental ecological principle that only recently is being applied to human industrial activities. A long history of integrating aquaculture and agriculture operations exist in many parts of the world. Polyculture, the growing of multiple species occupying different ecological or nutritional niches, is an example of a coupled-system approach to waste reutilization and minimization. A constructed wetland with sufficient capacity to stabilize the effluent from a linked aquaculture facility is another example of a coupled system.

Other coupled-system designs place the fish rearing unit within the context of a larger water impoundment and use the larger impoundment as an integral part of the waste treatment system. The use of rearing containers suspended or constructed in controlled water impoundment, as opposed to net pens in large lakes or the open ocean represents opportunities for efficient solids removal. Yoo discussed a system of raceways designed to float on top of a pond. Rectangular-shaped rearing units were attached to a dock structure that was suspended in a pond. Airlift pumps circulated pond water into the inflow of the raceway. A sediment capture device at the outflow served as the primary solids removal system. The water flow from the pond into the raceway was discharged back into the pond.

Brune experimented with a partitioned aquaculture system (PAS) in which raceway rearing units were coupled with high rate algal oxidation channels. The raceways, referred to as "fish confinement and solids removal units," were constructed 23 m long, 2.1 m wide, and 1.2 m deep to grow channel catfish. The last 2.6 meters of the raceway were recessed to form a sump designed to trap solids for removal. The outflow from the raceways entered the oxidation channels and eventually recycled back through the raceways. The growth of algae and its subsequent harvesting by filter feeding fish (Tilapia) was utilized to improve the water quality of the system.

Increasing integration between managed ecosystems and aquacultural operations may indeed be a necessary long term trend. However, it is difficult to envision a rapid changeover of an existing high intensity flow-through raceway production facility into a balanced ecosystem component, at least in the near term.

Flocculation and Chemical Filtration Aids

The joining together of individual suspended particulates into larger particles, flocculation, improves the efficiency of solids removal operations. Kinetic flocculation is an approach to particle flocculation without the use of chemicals. Flocculation is achieved by capitalizing on differences in particle settling velocities (faster ones overtake slower ones and coalesce with them), or by using velocity gradients within the water to cause particles in regions of higher velocity to overtake those in slower regions. The improved removal efficiencies in relation to the effort involved, however, are generally low and kinetic flocculation approaches are most often employed in conjunction with the use of chemical flocculating agents.

Flocculating agents, such as alum, are commonly used in the municipal wastewater treatment industry. A flocculating agent works to reduce the repulsive forces of particles suspended in water allowing them to clump together into "flocs," thus improving the effectiveness of gravitational separation techniques. However, the high costs, dosing concerns, and disposal issues associated with chemically enriched sludge have largely prevented the use of chemical filtration aids in aquaculture. Similarly, electro-coagulation, the use of a sacrificial electrode to introduce coagulating agents into the water, typically is not employed for solids removal in aquaculture. Recently, however, investigators at the Freshwater Institute in West Virginia have experimented with applying "natural" flocculating agents such as chitosan to rainbow trout (Oncorhynchus mykiss) recirculating systems in order to improve solids removal. Limited observations showed an improvement in water clarity, but a heretofore unknown toxicity effect also resulted in acute mortalities of the fish.

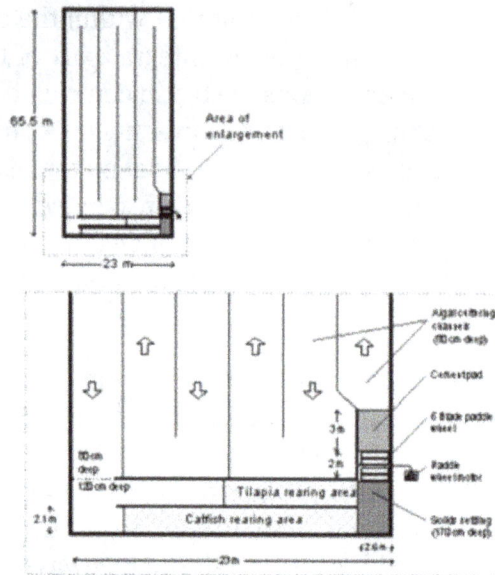

An experimental Partitioned Aquaculture System (PAS) consisting of an outdoor catfish production pond partitioned into fish production raceways and algal culture channels. The growth of algae and its subsequent harvesting by filter feeding fish (Tilapia) was utilized to improve the water quality of the system

Surface Filtration

Straining or surface filtration captures suspended particles by the use of a mesh-like material with a pore size that is smaller than the particles being removed. Various backwashing schemes have been developed to address the issue of screen "blinding" resulting from the buildup of captured particles and the growth of biological films. The Triangle Filter™ and various rotary screens are surface filtration devices that have found application in aquaculture. A conveyor filter is a variation of the rotating screen filter in which the filter surface is a wide belt of material that travels, conveyor belt style, around two axles. Water passes through the belt and the solids are captured thereon. Cripps (1994) reviewed the use of stationary screens, rotary screens (including axial and radial flow), chain type screens, and vibratory screens in aquaculture.

The fundamental difficulty with implementing surface filtration as the primary treatment system in flow-through raceway aquaculture is the result of the large volumes of water, with corresponding low concentrations of suspended solids that must be treated. Consequently, surface filtration devices in the United States have been installed primarily in aquacultural facilities with reduced water flows, such as those employing water recirculation.

Depth Filtration

Typified by sand filters and bead filters, depth filtration employs a medium in which the solids are captured via mechanisms of sedimentation, interception and diffusion. Although promising and employed in a variety of aquacultural situations, depth filtration of the effluent from the flow-through raceway culture of rainbow trout is typically not employed be- cause of the very large surface loading area that a media filter would require to accommodate the flow rates.

Gravity Sand Filtration

A gravity sand filter is often configured as a cement structure with under-drains. The structure is filled with a layer of sand 0.5 to 2 m deep. Wastewater enters the structure from the top and flows through the sand bed and out via the under-drains. Solids are retained within the media bed. When the filter becomes clogged with captured material the bed is taken out of service and back-washed by fluidizing the sand and re-suspending the captured material with high pressure water and/or compressed air. The re-suspended material is then drained from the basin as sludge.

The cost, including capital, operations and maintenance, of the filtration system is an important issue when considering granular-media filtration. The hydraulic loading rate of a sand filter is primarily determined by the surface area (length x width) of the filter bed. The surface area required by the filter bed is a major influence on the initial capital expense and operating costs. To treat the effluent from a high-intensity flow-through raceway (4,500 to 8,500 l/min) would require a sand filter with a surface area that is on the order of 40-70 percent of the surface area of the raceway. Assuming that all available land is occupied by production units, the farmer would have to take raceways out of production and convert them to sand filtration units to treat the discharge from the remaining raceways. Costs to the farmer would include the capital cost to convert an existing raceway into a sand filter, the backwashing equipment cost, the operational costs, and the opportunity cost of the lost production. Also adding to the total expense would be the fact that multiple filter beds would have to be maintained so that sufficient filtration capacity is available even when one bed is taken out of service for backwashing.

Pressurized Sand Filtration

The surface area requirements for sand filtration can be reduced by constructing an enclosed filter unit and forcing water through the sand bed under pressure. The use of a pressurized sand filter generally results in a longer filter run time between backwashing, and a reduction of the backwash water and time requirements. However, backwash savings have to be balanced by the increased pumping costs inherent in pressurized filtration. Pumping is a principal consumer of energy in most wastewater treatment facilities. Additional filtration capacity would be needed to accommodate the flow when one filter is taken out of service for backwashing. Wong presented calculations showing that the use of pressurized sand filtration to treat the water from the flow-through raceway culture of rainbow trout would represent about 70 to 110 percent of the annual feed costs.

Bead Filtration

Bead filters or expandable granular biofilters can provide both physical and biological filtration. These filters are well suited for mechanical backwashing methods, and have been installed in numerous recirculating aquacultural operations. Small plastic floating beads, approximately 3-5 mm in diameter, are loosely packed into a filtration chamber. Solids are trapped in the interstices of the beads as water flows up through the unit. The beads also provide a substrate for the growth of nitrifying bacteria. Solids are removed from the unit by agitating the beads with a propeller or by air scouring and then flushing the particulate-laden water. Potential disadvantages include high capital costs, higher head losses than sedimentation units, biofouling, water loss during backwashing and the inability to remove fine (< 20 micron) suspended solids. Bead filters are primarily used to recondition water in recirculating aquaculture systems and, similar to pressurized sand filters, are not well suited as a primary solids removal operation in flow-through facilities because of the large quantities of water that would need to be treated.

Gravitational Separation

Sedimentation, one of the oldest forms of water treatment employed by humans, essentially uses the force of gravity to concentrate suspended solids on the floor of a raceway or other basin. Its ease of implementation and relatively low operating costs has made the settling of suspended solids a widespread treatment operation in the raceway aquaculture industry.

A quiescent zone (QZ) is a sedimentation region typically consisting of the last one to five meters of the downstream end of a raceway. A screen that allows suspended particles to pass but excludes the fish is used to separate the upstream rearing section of the raceway from the QZ. The water flows through the length of the rearing unit and enters the quiescent zone where solid particles settle and deposit onto the floor.

Hansen discussed the use of an excavated depression in a stream bed to trap coarse-grain sediments, primarily of anthropogenic origin, that would be detrimental to trout habitat. Although conceptualized for trout habitat management, Hansen's "in-channel sedimentation basin" was, in essence, a precursor to the raceway QZ. In the 1980s, aquaculture facilities located on the Snake River in southern Idaho began experimenting with the use of screens to isolate the downstream section of the raceways to improve conditions for in-raceway sedimentation. Quiescent zones typically are cleaned by vacuuming, the wastewater being sent to off-line sedimentation ponds for further treatment. The modern QZ is, or should be, a carefully engineered structure designed to augment a raceway's natural tendency to accumulate solids. The use of QZs in raceways is considered a best management practice (BMP) and is recommended by the Idaho Division of Environmental Quality for rainbow trout raceway culture.

Increasing the effective surface area is one method to im- prove the capture efficiency of a QZ or any other sedimentation basin. Inclined tube and plate settlers have been utilized to enhance the performance of municipal wastewater sedimentation operations. Variations of these devices have also seen limited application in aquacultural systems. Cripps suggested that a "lamella separator," a type of inclined plate settler, had good prospects for use in the aquaculture industry, although initial capital expenditure would be high. A recurring issue with plate and tube settlers, both in municipal wastewater and aquacultural installations, is the accumulation of sludge and biological growth on the device. Procedures to address the solids accumulation problem, such as cleaning with a high pressure hose, can add substantially to the operating costs of these systems.

The solids collecting in a sedimentation basin typically arrive continuously, but are removed periodically, such as when the basin is cleaned. Settleable solids trapped in the sedimentation basin are still in contact with the water column. Unless removed from the system there will be continued breakdown of the solids resulting in smaller particles, solubilization of nutrients such as phosphorus, and an increased oxygen demand from biological activity. Based on observation of rainbow trout production operations, a wide variability in the cleaning intervals of QZs and other sedimentation basins exists. The movement of nutrients and other constituents from the solid to the dissolved state negates some of the benefits associated with rapid solids capture.

Static Water Aquaculture

Static water aquaculture is used for harvesting fish seasonally or perennially. The water supply for static water entities can come from a variety of sources, such as water run-off, sky ponds, natural water sources and from springs. The sustainment of a thriving aquaculture practice depends on the water quality. Making regular checks for monitoring quality, purity, potential overfeeding and amount of fertilization is vital. The quality of the soil that has been transported along with the water must also be controlled. The most desirable soil type for static water ponds is clay.

Static Freshwater Ponds

Science of freshwater pond fish-culture has made great strides in recent years and there is a fast advancing frontier of knowledge on every aspect of pond culture starting from farm designing and construction upto production of marketable fish of a wide variety of cultured fresh water species of finfish and shellfish. Examples are: carp culture systems in India, China, Israel, Germany, etc; catfish culture in U.S.A..

There is considerable competition with agriculture and other land-use agencies in this system of aquaculture and its success would, by and large, depend on comparative economics of land use. But much also depends on national policies on land use and the encouragement government gives to aquaculture as a means of producing fish protein.

Water quality should be the main concern for a successful pond aquaculture business. Regular checks for purity, quality, amount of fertilization and potential overfeeding that remains in the water column should be coducted.the soil that is being brought along with the water should also be controlled.

Another important factor for a sustainable business is the soil quality and the soil that is being brought along with the water should also be controlled. The soil which is mostly chosen for its ability to hold the water in the pond is clay.

Chapter 3

Fish Farming and Fisheries

Fish are commercially raised in tanks or fish ponds in fish farming. An important entity involved in raising and harvesting fish is a fishery from which wild fish, farmed fish, fresh water or saltwater fish may be captured. Fisheries can be freshwater or saltwater. This chapter elucidates the fundamental aspects of intensive and extensive aquaculture, fish farming types, issues, indoor fish farming, etc.

Fish Farming

Fish farming refers to the commercial production of fish in an enclosure or, when located in a body of freshwater or marine water, in an area that is penned off from the surrounding water by cages or open nets.

A fish farm is similar to a fish hatchery in that both can contain 500,000 and more fish. But, a fish hatchery is designed to raise the fish only to a young age before they are released into the wild, usually to bolster the numbers of that species. In contrast, a fish farm is designed to raise the fish until they are a size and age that makes them the best commercial value. The fish are ultimately retrieved and sold, typically as whole or processed food.

Fish farming is the most common form of aquaculture, and commonly involves trout, salmon, tilapia, cod, carp, and catfish. For a species such as cod, whose numbers in the Grand Banks fishery off the east coast of the Canadian Maritime Provinces plummeted to near zero in the 1970s due to overfishing, and as of 2008 have yet to recover, the cod available from fish farming represents almost the sole source of the fish in North American markets.

The example of cod is cited as one of the advantages of fish farming. Raising fish under more controlled conditions that are possible in the wild avoids the problem of overfishing. As well, because an operation takes up relatively little space, feeding and care of the fish can be done under more controlled conditions, which is an economic advantage to those who own and run the facility.

However, fish farming is a controversial practice. For example, on the Canadian west coast, the farming of salmon typically uses species normally found in the Atlantic Ocean. The escape of fish to the wild does occur, and has created concern that the presence of the species in an environment that is unnatural to them could upset the marine ecology. Other concerns of fish farming are the overcrowding of fish, which can make them more susceptible to disease such as sea lice, and the use of antibiotics, which can also be released into the natural environment.

Fish farming is an ancient practice, dating back to about 2500 BC in China, when carp were raised in ponds and in artificial lakes created by receding floodwaters. Some of the motivations for fish farming in ancient China are shared with fish farm owners and operators in 2008. These include maximizing the food available from the resource; reducing the energy needed to search for, gather, and transport the food; making food production more predictable and less likely to be influenced by weather, predators, or other factors; and ensuring that the quality of the resource remains acceptable over time.

Evidence of fish farming also dates back at least 1,000 years in Hawaii, when rocks were added to existing reefs to create an artificial pond. The spaces between the rocks were large enough to let seawater circulate in and out, allowing nutrients to circulate in and waste out, but were too small to allow the fish to escape. The open net design of present-day fish farms follows this example.

Fish farming became more prevalent in Europe in the fifteenth century. The first known fish hatchery constructed in North America was built in the Canadian province of Newfoundland in 1889.

In the 1960s, fish farming expanded worldwide as some commercial fish stocks became less plentiful and a growing global population increased the demand for fish. As with factory farms—the landlocked facilities in which huge numbers of poultry and livestock are raised—economic incentives were provided to encourage the establishment of freshwater and marine water fish farms. In addition, corporations involved in the sales of fresh and processed fish products began to expand into fish farming as a way of ensuring supply, expanding their market, and trimming costs.

As practiced in China thousands of years ago, fish farming was efficient and sustainable. The numbers of fish were suitable for the space available and the population was managed so that the numbers of fish ready for harvesting did not decline over time. When done in such a sustainable way, fish farming can be a good strategy to supplement or even replace the fish caught from the wild.

However, the confinement of large numbers of fish in a small area can create problems. In an enclosed pond, one problem can be the accumulation of waste products. Aside from making the water less hospitable for the fish, the waste material can act as a food source for microorganisms known as algae. Combined with suitable water temperature and sunlight, the presence of the food source can lead to the rapid increase in the number of algae termed as algal blooms. The number of algae in blooms that occur in the open ocean can be so large that the growth is visible from orbiting satellites. In a confined pond, an algal bloom can use up much of the oxygen in the water, leading to the death of the fish.

When a fish farm is done in tanks (closed circulation type), it is essential to keep the water well oxygenated and to remove wastes. Bubbling air into the water as is done with a home aquarium can be one means of oxygenation. Alternatively, water can be cascaded from tank to tank in the fish farm complex, with oxygen being supplied as the water tumbles between tanks. Waste removal is typically done by continual removal of the water and passing it through a filter before returning the water to the tank. Regular monitoring of the tank water is necessary to ensure that other parameters such as pH are maintained at an optimal level.

Fish farming that is done in large bodies of freshwater or in the ocean avoids the potential problems associated with the closed circulation system. Typically, the fish are housed in a series of pens, which consist of mesh nets attached to rigid supports. The entire structure floats on the surface.

Impacts and Issues

Fish farming has become a very contentious practice, for a number of environmental reasons and for the adverse health effects it has on the farmed fish and possibly other species, including humans.

In a fish farm, the concentration of fish far exceeds that found in schools of fish in the wild—50,000 or more fish in an area of several acres in volume—with the possible exception of the spawning runs of west coast salmon. These crowded conditions reduce the free-swimming volume of each fish to about that of the average household bathtub. In such crowded conditions, the fish bump and rub against each other in the boundaries of the pens, which can produce cuts and scrapes. This increases the likelihood of infection and the development of diseases.

Species of sea lice that parasitize Coho and Atlantic salmon are especially troublesome. The sea lice attach to the fish and feed on tissue, which creates lesions and causes fluid loss from the affected fish. The confined fish become ill and can die. In addition, the sea lice can spread to wild salmon in the seas around fish farms when farmed salmon escape from the confinement, and also when the lice are washed away from the fish farm into the surrounding water. A 2001 survey of wild juvenile salmon migrating past fish farms in British Columbia found many more sea lice on the juveniles that had passed the farms than on those who had not yet passed by the facilities.

The escape of fish from fish farms is not a trivial and isolated event. Rips and breaks in the pen material and buffeting of the pens by storm-driven waves can lead to the escape of fish. In some cases, pens are designed with a net lid to reduce this possibility. Sometimes only a few fish escape. But mass escapes have occurred. For example, in January 2002, over 8,000 fish escaped from a fish farm in Clayoquot Sound, British Columbia. Worldwide in 2004, an estimated 2 million farmed fish escaped to the wild.

Once in the wild, the farmed fish have the potential to transfer disease to the wild population. An article in a December 2007 issue of Science documented declines in the population of wild Pacific salmon related to their decimation from sea lice transferred from farm populations of Atlantic salmon. The situation is so dire that the natural population could be reduced by 99% by 2015, which would be an economic disaster for the traditional salmon fishery and those employed by the fishery.

Antibiotics can be supplied to the food in an effort to control infections. As with land-bound factory farms, this practice encourages the development of antibiotic resistance among the surviving bacteria. These hardier varieties of the bacteria may pose a health hazard not only to the farmed fish, but to wild fish populations and to humans.

A fish farm releases a great amount of untreated sewage to the surrounding water. A study done in Clayoquot Sound calculated that the 700,000 fish housed in the facility that is the size of three football fields generate the daily equivalent amount of sewage produced by 150,000 people.

Extensive Aquaculture

Extensive aquaculture includes systems of culture and rearing in which human intervention is concentrated on the reproduction of the stock, in addition to capture. As compared to fishing, these systems make it possible to selectively increase the production of the species most useful for man's food (salmon, oysters, etc.), industry (algoculture), or pleasure (sport fishing, pearl culture, etc.). They use the planktonic or benthic production of the water masses, thus economising on the fertilisation of the breeding environment (semi-intensive systems), or the fattening of the stock (intensive systems).

Extensive practices are not new. In Europe, the fattening of mussels on stakes dates back to the 14th century. The value of capturing spat as a remedy for the overexploitation of natural oyster beds was realised as early as the second half of the 19th century. The value of transplanting young plaice from their nurseries in the south of the North Sea to less densely populated lagoon or sea fattening zones, at the beginning of this century. The mastery of artificial reproduction stimulated enthusiasm for sea ranching: the first releases of young salmon from hatcheries date from the middle of the last century, the first cod hatchery was built in Norway in 1884, and the first lobster hatchery in 1891 on the Atlantic coast of Canada.

Knowledge about the actual production of several extensive systems is still limited. While the production of sedentary species (shellfish and algae), whose stocks are geographically distinct from the natural beds and generally owned by an individual or a corporate body, is fairly well known, the culture of vagile species (mainly fish and crustaceans) whose stocks are exploited collectively at the same time as wild stocks and whose captures are not distinguished from those of the natural populations, is not at all well known. This uncertainty has led the FAO to exclude from its aquacultural statistics the production of collectively exploited stocks. This decision leads to the inclusion in the fisheries statistics of production resulting from the release of young fish into water courses, lakes, coastal waters, and the ocean, even though this production is very significant for several species, in particular salmon.

Algoculture and shellfish culture, with a production of 7,7 and 7,3 million tonnes respectively in 1996, based almost entirely on the extensive modes, represent half of world aquacultural production as defined by the FAO. European shellfish production exceeds 750,000 tonnes. In Japan, the production of chum salmon by ranching has increased tenfold over the past thirty years, to exceed 200,000 tonnes today. In 1994, the number of migrating fish in the rivers of Hokkaido was ten times the previous maximum observed in the entire history of fishing.

From the ecological standpoint, the potential for the expansion of extensive aquaculture appears considerable. Theories on the reproduction strategies of aquatic populations, the amplitude of the natural fluctuations of wild populations, and the productions achieved in certain extensive farming systems indicate that aquatic ecosystem could support stocks far bigger than the average biomass of the wild stocks. Furthermore, since the volume of the marine ecosystems exploited by these methods is still relatively small, it is understandable that this mode of production should stimulate research interest. The number of marine species studied rose from 150 in 1984 to 250 in 1994.

The results have been disappointing however. In recent years, few R&D programmes have led to any significant production. Shellfish culture is experiencing overstocking problems which are resulting in the deterioration of the technical and economic performances of the enterprises concerned. In the effort to reduce these imbalances and manage the conflicts which result from the competing uses of coastal ecosystems, traditional methods of regulation have proved ineffective.

The prospects for extensive aquaculture can be evaluated with reference to the state of the sector. Since the 70s, the annual rate of increase in world halieutic production has constantly been falling - in value terms even more rapidly than in tonnage -, to become negative in the course of the this decade. The reason for this decline is the exhaustion of wild stocks and the ineffectiveness of the traditional methods of regulating access to fisheries whose possibilities for expansion are exhausted. The aquaculture development policies introduced at the end of the 60s to take over from conventional fishing started to bear fruit during the 80s. For the past fifteen years world aquacultural production has been growing at an annual rate of 8.5 %. It now accounts for over a quarter of total production. Progress remains unevenly distributed however in terms of cultivation systems, production regions, products and markets. Despite the spectacular success of salmon farming in northern Europe or prawn culture in the tropics, traditional systems (small shellfish farms, fresh-water fish farms, and algoculture, mainly in Asia) remain the main beneficiaries of the new expansion. Due to the stagnation of fishing catches, European countries as a whole are faced with both a quantitative and qualitative shortage of sea products. In 1996, they imported over US$ 21 billion of sea products, and the deficit amounted to US$ 6 billion. For the European Union alone, it was US$ 8 billion - and respectively US$ 16 billion and US$ 4 billion for Japan and the USA. Only Norway, Iceland, Russia, Denmark, Ireland and Ukraine are, in descending order, net exporters in value terms. As in most of the rich countries, the shortage was mainly in high quality products (noble fish, crustaceans and shellfish, fresh, filets and frozen). In these conditions, public policies should continue to support the expansion of aquaculture and the diversification of production systems, with priority being given to those concerned with species of high commercial value.

The marine environment is well suited for the analysis of the implications and possibilities of extensive aquaculture. Because of their large dimensions, and because, even in coastal areas, the seas and oceans suffer on average less pressures than inland waters, marine ecosystems offer better prospects. The scale of the oceanic masses reveals better the problems that the fluidity of the milieu and the mobility of the stocks represent for their domestication. Any advance in the law of the sea sheds better light on the role of the institutions in the exploitation of natural resources and the rationalisation of their uses.

Main Extensive Systems

Process of Domestication

Technical intensification makes it possible to exceed the productivity limits of the wild stocks. The process consists of extending the technical controls of physiological functions such as reproduction, particular stages of the life cycle (eggs and larvae, juveniles, adults, reproducers), as well as their cultivation milieus. This domestication takes place through discrete jumps: between fishing and completely controlled cultivation there is a range of systems that can be classified in four groups: (i) those which concern essentially reproduction, (ii) those which also intervene in the fertility of the cultivation milieu, (iii) those in which the stock is also fed, and (iv) lastly, those in which the wastes are also recycled. Not all of the technically conceivable systems are viable, and the most intensive are not necessarily the most efficient.

While technical intensification reduces the role of the natural resource, it does not eliminate the dependence of production vis-à-vis the environment. It changes only the nature of the constraining factors and the limits that they impose on production. In fishing, it is the tertiary production - that of the halieutic populations - which is the limiting factor. In aquacultural systems, the stock, which is artificially reproduced, is no longer limiting. It no longer constitutes the natural resource. Appropriated, it becomes assimilable to capital. In extensive cultivation it is the primary and secondary productions (plankton and benthos) serving as food for the stocks which limit production. In intensive systems where the food is added it is the wastes (unconsumed food, faeces, antibiotics and medicines, etc.), as well as escapes of individuals who may modify the genetic heritage of the wild populations, that need to be adjusted to the acceptance capacity of the aquatic ecosystem. Throughout the process of domestication, the capital and labour intensity of the operation increase at the same time as the technical intensification. The increase in productivity (per unit of area) thus obtained is accompanied by an increase in the pressure on the environment.

In the course of this process the economic and social organisation of the operation also changes. As the technical controls diversify, the rules which define and support their exercise - and in particular the regimes governing the ownership of the natural resource - change. Mazoyer and Roudart thus describe the links which have developed in agriculture between the domestication of the natural resources, the intensification of their exploitation, and the land ownership regime: 'It emerges from this long history that, since the Neolithic age, the "ownership" of the land has gradually extended to the different categories of land as they have become artificialised: built on land first of all, then gardens and enclosures enriched and cultivated each year, cleared land bearing crops, developed hay meadows, land cultivated between two uncultivated or fallow periods, land cultivated continuously, improved pastures, developed and maintained forests. The ancient rights of common use (hunting, gathering, gleaning, use of the common, right to collect wood), for their part, have always concerned land where, without any particular work, wood, natural grasses, the wild grasses of fallow land, game, etc. continued to develop spontaneously'. The history of shellfish culture in Yerseke in the Netherlands shows that the processes in aquaculture do not differ fundamentally from those observed in agriculture. In the course of the process of technical intensification, the structures of the production, exploitation and regulation systems were transformed in stages:

- Before 1860: collective exploitation of natural beds by small local fishermen; regulation of access by the administration;

- From 1860 to 1933: with differences between oyster culture and mussel culture, transition from fishing to extensive aquaculture, with the public auction of individual concessions in the public maritime field, and privatisation of the stocks; application by shellfish culture cooperatives of individual production quotas aimed at regulating exports;

- 1933 - 1967: intensification of production; epizootic diseases; administrative regulation of production and markets;

- After 1967: vertical integration of enterprises, which become capitalistic; reduction of public intervention.

Description of Extensive Aquaculture

Extensive aquaculture is characterised by a production function in which human intervention is concentrated on the reproduction of the stock. This intervention may take two forms:

- Through the physical creation of particular habitats: seasonal closing of lagoons to hold the juveniles before their migration to the sea (valliculture, which includes other interventions such as the regulation of the water circulation, the elimination of undesirable species, or the provision of additional food, etc.), creation of fry ponds and artificial nurseries to support the reproduction of natural populations (salmon), capture of natural spat by providing substrata for the collection of larvae (shellfish culture), creation of niches for the juveniles of certain species when the density of the former is a limiting factor, etc.

- By releasing hatchery fry into the natural milieu.

Representation of Extensive Aquacultural Systems

The first set of examples shows that artificial reproduction is not essential to start a new system. On the other hand, the capacity to act on the reproduction of a population through forcing its recruitment is.

Extensive systems aim at three objectives:

- Reconstitution by restocking of natural populations reduced by the deterioration of their environment, in particular the degradation of habitats which are critical for reproduction (pollution, dams on water courses which impede the migration of reproducers, disappearance or silting up of the fry ponds and nurseries of anadromous fish, etc.);

- Acclimation of exotic species or populations in order to create new stocks more valuable than the native stocks;

- Simple enlargement of a stock by releasing young fish into the natural milieu, often resulting in overpopulation.

As the stock feeds on the natural milieu, extensive fish farming is sometimes qualified as 'production', as opposed to intensive fish farming, qualified as 'processing' in which artificial feeding is provided. In the sea, semi-intensive systems in which at least part of the food is produced in the cultivation milieu do not exist (outside lagoons). This is not the case in agriculture or fresh-water fish culture. This difference is explained by the fluidity of the aquatic milieu and the scale of the water masses which limit the possibilities of fertilisation to units of sufficiently reduced dimensions - hence inland (ponds, lakes and lagoons) -, and also by the state of ownership regimes in the maritime field. In the sea, the ownership of stocks is limited virtually exclusively to sessile or sedentary species (shellfish, algae), and vagile species (fish) in captivity (fish culture in cages or basins). The importance of intensive cultivation in the marine field is perhaps due as much to the need for the appropriation of the stock as to the enhanced zootechnical performance.

Extensive systems which aim at fattening naturally result in overpopulation (as compared with the biomass observed under natural conditions). The increase in biomass – and hence finally in the total production of the stock - is limited by the trophic capacity of the ecosystem used as the cultivation milieu. The growing tension between the biomass in expansion and the limited productivity of the cultivation milieu results in a regular decline in yields (decline in growth performance, increased mortality and increased risk of epizootic diseases).

Extensive cultivation does not modify the environment very much. However, the employment of selected strains for the purposes of cultivation or exotic strains for the rearing of young fish risks impairing the genetic heritage of the wild populations living in the same ecosystems.

Recapture is facilitated when the stocks are made up of sessile or sedentary species, or when fishing is carried out close to the coast or in rivers where the stock concentrates periodically (salmon during their reproduction, for example).

Like halieutic stocks and intensive cultivation, extensive aquaculture is subject to the effects of pollution by industrial, agricultural and domestic wastes. The fluidity and density of aquatic milieus makes them more sensitive than land and atmospheric milieus. Aquatic milieus are in fact better carriers than land and atmospheric milieus, and leaching and transport by water course networks concentrate the greater part of the residues from human activities in the catchment areas in the rivers and coastal areas. On the world scale it is estimated that some 70 % pollution ends up in the sea.

Physical Development of Habitats

A halieutic population can be supported by the development of habitats critical for its reproduction. This is the case, for example, with ladders to enable salmon to get past dams on the rivers they use for their reproduction migration, arrangements to conserve the flow and quality of river water, or the maintenance of their fry ponds.

Artificial reefs are the most common example of physical measures in the marine field. It is in Japan, where it is planned that they will cover 12 % of the coastal areas by the year 2000, and along the Mediterranean coasts in Europe, that these arrangements are the most developed.

Their promoters attribute the following virtues to artificial reefs:

- Concentration of the fish by tropism (provision of food and protection against predators);

- Increase in biotic diversity, for example when the reefs are installed in sedimentary zones poor in sessile and sedentary species (algae, shellfish, urchins, crustaceans, rocky ground fish, etc.);

- Increase in the biomass, effectively observed in the case of sedentary and sessile species, but not so obvious in the case of vagile species (fish);

- Physical elimination of tackle such as trawl nets.

Depending on the objective, the species it is desired to support, and the fishing methods it is intended to favour, the materials and architecture of artificial habitats vary considerably.

In the creation of artificial reefs, the exclusion of non-indigenous species by residents preoccupied by the rarefaction of natural resources in the sectors that they are accustomed to frequent is often an important motive, even though it is not always expressed. This monopolising of space may facilitate initiatives in favour of conservation. In the USA, for example, artificial reefs are used to increase the floral and faunal richness and promote the development of ecological tourism. The example of floating rafts and other arrangements to concentrate fish shows that these constructions can effectively attract fish. Through bringing the fishing grounds closer, stabilising the presence of the fish and facilitating the spatial allocation of fishing sites, artificial reefs can aid fishing and its regulation. But the same results can also be achieved by other methods, possibly less costly (floating rafts), or more direct (shellfish culture).

The effect of artificial reefs on the production of halieutic stocks remains to be demonstrated. Up to now, the observations have virtually exclusively been concerned with spot densities of adults, whereas to demonstrate the impact of the reefs on halieutic productivity it would be necessary to be able to measure their effect on the absolute abundance of the populations. This implies studying the process of regulation, on the scale of the population, and at the level of reproduction and the early stages where these processes act. While such studies may be complex and costly, they are nevertheless realisable. Only such investigations would make it possible to establish whether artificial reefs are the best way of achieving the objectives assigned to projects for the implantation of reefs and, if this is the case, the conditions that must be fulfilled (nature, localisation, extension) to obtain the best impact. Artificial reefs, for example, are likely to effectively support the recruitment of halieutic stocks provided that they cover a sufficient proportion of the coastal nurseries. On the other hand, through acting on the effects (stocks) and not on the causes (regulation of access) of overfishing, they can at best retard, but not prevent, overexploitation.

Shellfish Culture

In shellfish culture, the forcing of recruitment is achieved in different ways:

- By Simply taking juveniles from the natural beds and replacing them on the bottom or in the water on artificial supports (hurdles or nets) for fattening, - i.e. in milieus that the larvae cannot colonise under natural conditions; this technique is widely used in mussel culture;

- By capturing spat on artificial supports placed in the natural milieu; the larvae which attach themselves on the collectors come from natural beds and, above all, from cultivated stocks once the system is well established; as with the preceding technique, the spat is then transferred to the fattening milieu (on the bottom or in suspension); this technique is that most generally used in oyster culture;

- By the production of juveniles in a hatchery (scallops, oysters, abalone, Philippines clams (Ruditappes philippinarum), etc.).

The more or less complete mastery of recruitment has permitted three major advances:

- Creation of stocks whose biomass may considerably exceed that of the natural populations; thus, in the Marennes-Oléron basin (France), the oyster biomass now regularly reaches 200 000 tonnes, or several times the highest ever reached by the natural beds;

- Freedom from the climatic hazards of natural recruitment and their effects on annual production; again for the Marennes-Oléron basin, the capture of spat has been zero for only two years out of 45 years of observation for the Portuguese oyster, and low or zero only three years out of thirteen for the Japanese oyster (Héral et al., 1986);

- The extension of production to sectors where the species does not reproduce; this is how in France, the cultivation of the Japanese oyster north of the Loire is achieved thanks to spat captured south of this river.

Thus, a rustic production chain, not initially requiring the complete mastery of the biological cycle, has turned out to be economically efficient and well adapted to the exploitation of particularly productive coastal ecosystems. This development has taken advantage of three particularities of shellfish ecology:

- Their sedentariness which facilitates the confinement and appropriation of the stock;

- Feeding by filtration which makes it possible to exploit the most productive lower levels of the food chain and to achieve a high concentration of cultivation;

- Their good tolerance of climatic variations.

But the take-off of the system was achieved only because these technical advances were complemented by institutional innovations:

- Privatisation Of The Stocks: Made Physically Possible By The Sedentariness Of The Species, The Possession of the stocks has been formalised by the attribution to oyster growers of private concessions on the public maritime domain; having the necessary guarantees for their investments, the growers have been able to invest in the reproduction of the stock;

- Intervention by the public authorities in the conservation of the quality of the oyster farming ecosystem, the control of the health of the stocks and the health quality of the products.

Shellfish culture has been developed above all in Asia (China, Korea, Japan, etc.) and Europe. In France, for example, the shellfish culture turnover amounts to over one-third that of fishing and employment is one and a half times that of fishing. The big differences seen in European production, both between countries and between species, suggest that the constraints which impede the development of shellfish culture are social and institutional rather than ecological or technical. There should therefore be possibilities for expansion in countries where production is still modest.

90% of European production is accounted for by oysters and mussels. While both oyster and mussel growing have been modernised in recent years, Europe has not achieved any successes comparable to that of Asia with the development of the cultivation of the Japanese scallop (Pecten yessoensis). Mastered in Japan at the beginning of the 50s, production reached a million tonnes in China and 265 000 tonnes in Japan in 1996. There are nevertheless prospects for diversification in Europe. The cultivation techniques of several species have been mastered, but the take-off of commercial production comes up against various difficulties: ecological and economic for the abalone (Haliotis tuberculatus), epizootic (brown ring disease) for the Philippines clam (Ruditappes philippinarum), and legal for the scallop.

Sea Ranching

While, as with the sedentary species, the fattening of vagile species is possible in the marine milieu, their mobility seriously complicates the appropriation of the stocks and the captures. The systematic marking of hatchery products is made difficult by the small size of the individuals. Above all, it is difficult to ensure that the ownership of the stocks is respected at sea because of the remoteness of the fishing areas and of the low density of boats on them, as well as the transitory nature of the catches. This is why the ownership of stocks of vagile species is practically never acknowledged at sea. Contrary to what happens with livestock in agriculture, fish that escape from private farms become public property.

For certain species these difficulties have been partially overcome by the exploitation of a trait common to many aquatic populations, the phenomenon of homing. Through choosing species which congregate to spawn in coastal or inland waters where access to fishing is restricted and easier to control, enough of the stock can be recaptured to cover the cost of rearing the fry. Sea ranching is based on this strategy. Technical success is assured if the weight gain in the individuals recaptured during the fattening phase exceeds the combined losses due to natural mortality, straying – not all individuals return to the point where they are released -, and catches by third party fleets. When these conditions are fulfilled sea ranching offers the possibility of harnessing the unexploited primary and secondary productivities of vast maritime areas.

The risk of interception by third party fleets increases, on the other hand, with the area and duration of the freedom phase. The salmon ranching programmes of the Baltic countries (mainly Sweden and Finland) have been so successful technically, for example, that they have given rise to the development of plurinational fishing at sea. The captures have reached such a level (some 3,000 tonnes, or 80 to 90 % of the annual recruitment) that they are now threatening the wild stocks fished at the same time. Through ensuring the profitability of fishing, a rearing programme intended to conserve the natural populations decimated by hydraulic schemes and river-borne pollution is now threatening their survival. The answer would be to prohibit fishing at sea. This is the solution adopted in the North Pacific where the states have agreed to prohibit salmon fishing on the high seas, or in Iceland where salmon fishing has been prohibited in coastal waters for the past 60 years.

The system was first developed for salmon. For these species in fact, ranching offered a solution to the crucial problem of the conservation of populations threatened by the degradation of their inland water habitat on which they depend for their reproduction and in which they are particularly vulnerable. Salmon exhibit a number of traits that favour this type of enterprise:

- they reproduce in very localised fry beds in water courses, where their recapture is easy, after a phase of oceanic fattening in the course of which the individuals acquire over 90% of their final weight;

- they have a very marked homing behaviour, with a generally low rate of straying;

- the particularities of their demographic dynamic (the cohorts reach their maximum biomass at the moment of reproduction, after which they all die - Atlantic salmon - or most of them die - Pacific species) theoretically make it possible, through short-circuiting the freshwater phase, to compensate for the problems with the conservation of inland waters and the halieutic stocks; again in theory, the regulation of their fishing could be replaced by simple management of the annual releases;

- artificial reproduction is easy to achieve (salmon are one of the small number of aquatic species which lay large eggs), and was mastered very early;

- the species is of high value for human food, or even more for sport fishing.

For a long time the impact of the experiments carried out in the Atlantic and North Pacific was nil or not proven. It was not until the 60s that, simultaneously in Sweden and the USA, the technical viability of the method was able to be established. Since then, advances in knowledge have revealed the critical importance of factors which had been neglected for a long time, such as the marking and identification of hatchery individuals in catches for the evaluation of survival rates and conditions, or the risk of genetic mixing or pathological contamination with wild populations, etc. Since then the list of first technical and then economic successes has lengthened. In the literature, the turbot in Europe, and the gilt-head bream (Pagrus major), the flounder (Paralichthys olivaceus), the chum salmon (though perhaps no longer since the development of fish farming in cages has brought prices down) and the Japanese prawn (Penæus Japanicus) in Japan (Kitada, op. cit.), are cited among the species whose production by ranching is considered economically viable. On the other hand, in Europe, research into lobster (France, United Kingdom, Norway) and cod (Norwegian PUSH8 programme) reached negative conclusions.

The interest of sea ranching is not limited to the conservation of marine fauna. The technique may meet three objectives:

- Restocking for the conservation of natural populations. This is the most common case. Thus, when Sweden launched its hydroelectric programme, it imposed on its electricity company a compensatory fish rearing programme. The other Baltic countries subsequently introduced similar measures; in the second half of the 80s, these countries released over 6 million smolts each year. Whereas only 20 Swedish rivers out of 60 are now accessible to salmon, and natural recruitment has fallen by 90%, the resource is quantitatively conserved;

- Overpopulation to support commercial and recreational fishing: in Japan, the releases of chum salmon amount to 2 billion individuals a year, and the recapture rate increased from 2% in 1963 to 15% in 1995; on the East Coast of the Pacific, a decade ago salmon ranching provided 20 % of the total Canadian salmon catch, and one-third of the pink salmon caught in Alaska; but, as far as can be judged given the disparity of the statistics, on the world and European scales sea ranching production remains very much lower than that of shellfish culture or algoculture;

- Acclimation: there have been several instances of the successful introduction of non-indigenous species: chinook salmon in the rivers of the South Island of New Zealand; chinook and coho in Chile; coho in New Hampshire; but the results of these transplantations are often not very convincing, as in many cases the stocks can only maintained through repeated restocking with young fish, so that in the end these become simply natural milieu fattening programmes. They may be perfectly profitable however, as was concluded with the experiments carried out in the Kerguelen Islands in the Southern Ocean, for example.

The long history of transplantation of young plaice in the North Sea is worthy of mention. Launched in 1891 in Denmark and pursued for over a century, the programme has been extended to other species (cod, salmon, trout, whitefish, turbot and eel), and repeated in other countries (Germany, United Kingdom, Norway and Sweden). Although their technical efficiency has been demonstrated and economic profitability has been achieved in several configurations, these programmes have never given rise to significant commercial development. The reasons for this relative failure are political and financial. In Denmark, for example, the programme gives rise to disputes over the ownership of the stock. Fishermen of the coastal regions of the North Sea where the juveniles are fished claim they should conserve rights over the transplanted stock, just as the Limfjord fishermen, even though no professional fisherman contributes to the financing of the transplantation. Only the sports fishermen of the Limfjord pay a fee, and this represents only a small proportion of the total cost. In these circumstances, no significant private financing is going to take over from the public funding which for its part remains modest.

The organisation and management of ranching programmes differs according to the objectives. Programmes which are limited to the conservation of wild populations, without concern for profitability, are most often implemented by public bodies, even though the production of the young fish is sometimes subcontracted to private fish farms. The creation of significant stocks involves two questions: the reservation of their exploitation to particular groups (groups of national and local fishermen, professionals or anglers), and the contribution of the main beneficiaries of the catches to the financing of the breeding farms. The state of Alaska, for example, grants licences to private hatcheries for non-profit operations which are managed by local cooperatives and communities of native fishermen exploiting the stock. Fishermen can pay a fee for the financing of breeding programmes. This is the case in the United Kingdom, where the landowners finance fish rearing programmes and then rent the fishing rights on their reaches to anglers. In Japan, professional fishermen pay a contribution calculated on the basis of the sales price of the fish (3 to 5%) to finance coastal breeding farms, while access to the coastal fisheries requires belonging to a cooperative or a group of private fishermen. Certain countries (Iceland, Chile and, in the USA, the states of Alaska, Oregon, Washington, and California on an experimental basis, etc.) authorise the creation of farms whose aim is private production for profit. In most cases however, private ranching projects come up against the opposition of both professional and sport fishermen, as well as the general public, who object to the calling into question of public rights, the private exploitation of public resources, or the risk to professional fishing of a fall in fish prices.

Systems for Retaining Fish in Open Milieus or very Large Enclosures

In order to limit the losses due to the straying of fish and catches by third party fleets, attempts have been made to reduce the area of straying of the stock through conditioning juveniles before their release by emitting a sound signal during feeding. This training may facilitate the recapture of the fish. It is in Japan that experimentation is most advanced (notably for the gilt-head bream - Pagrus major). These trials started in the 70s, but have not proved very conclusive. The fish get used to the signal and gradually move away from its source. Though the idea is intellectually attractive, the method involves several unknowns which need to be clarified for each species and on each site:

- Conditions and cost of acquisition and conservation of the behaviour (control of habituation); if the natural food has to be permanently complemented then ranching looses its interest;

- Risk and cost of conditioning undesirable species (predators and commensals);

- Coherence between the spatial scale of the project, the economic and social structure of the operation (public, collective or individual), and the ownership regime of the stock and of the colonised ecosystem;

- Distribution of the costs and benefits between the protagonists of the project, the exploiters of the stock, and the other users of the ecosystem and of the space.

In the present state of knowledge, the method appears to be essentially a refinement of already mastered systems. Focusing on technical innovation, it under-estimates the institutional reforms on which its success depends.

Stocks are also confined, or it is planned to do so, in large natural spaces (lagoons, bays) by physical means (dams, nets, electric screens, bubble screens, etc.). This method also underlies the valliculture traditionally practised in Mediterranean lagoons. It is very developed in Asia, traditionally in fresh waters, more recently in coastal zones. As compared with physical barriers, bubble screens or electric screens has the advantage of not being subject to fouling or the accumulation of debris (algae), nor to damage by bad weather. On the other hand, their effectiveness remains to be proven. Some people consider this to be the way forward for aquaculture, in so far as the method would permit the extension towards the sea of intensive systems whose development in the coastal zone is blocked by constraints of space and conflicts of use, or is likely to be in the near future.

Conditions Required for the Take-off of New Extensive Systems

Production of Quality Juveniles

Although the small size of the eggs of the great majority of aquatic species complicates artificial reproduction, hatchery techniques have improved considerably over the past thirty years. In aquacultural research, the mastery of this function has often been considered a priority, even if artificial reproduction is not always essential for startup extensive cultivation (cf. shellfish culture or the transplantation of young plaice). Today the range of species whose reproduction is well or fairly well mastered is relatively large and given present knowledge and know-how this list could be fairly rapidly extended in the future. The production of juveniles is thus not a major obstacle to the development of extensive aquaculture.

On the other hand, the interest of artificial reproduction (uniformity of the product, reduction of natural hazards, less dependence on seasons and genetic selection) grows with the development of extensive systems. Even if the potential applications are less diversified than in intensive cultivation, genetic selection may contribute to the fight against epizootic diseases, improve growth performance, or serve to mark the individuals released (to help evaluate survival rates or genetic mixing with wild stocks). But selection also creates risks for the integrity of the genetic heritage of natural populations. At present, the lack of basic knowledge about the genetics of certain taxonomic groups, such as molluscs, is a constraint for selection.

Forcing Recruitment

For most of the species which have been the subject of prolonged experimentation there is an abundant literature describing:

- The qualities that the released individuals must possess: origin of the stock (natural or hatchery juveniles, genetic heritage as a function of the ecosystem to be colonised), size and age, adaptation to the life in the natural habitat (in the case of salmon, for example, prior acquisition of the reflexes and behaviours of feeding in the wild and defence against predators);

- The conditions for the release of individuals: habitat, date with respect to the smolt stage for salmon, seasonal trophic enrichment of the milieu, occurrence of predators, etc., with, as a possible corollary, operations to prepare the site (elimination of predators such as starfish in scallop cultivation in Japan, for example).

But, even for a single species, the conclusions are not always very concordant. This is explained by the essentially zootechnical approach often preferred in R&D programmes. In the absence of appropriate knowledge on the ecology of juveniles and the reproduction strategies of populations in their environment, it is difficult to define the qualities that the alevins have to exhibit and the conditions necessary for their release.

The constraints of aquatic life - and in particular dispersion - impose complex demographic strategies on marine populations. These strategies exhibit a number of common traits however:

- Very high fecundity, to compensate for the enormous mortality suffered by the eggs and larvae whose means of locomotion and food reserves are limited;

- The dependence of each population on a succession of hydrodynamic structures and the selection of physiological processes and behaviours adapted to the conditions obtaining in the structures occupied in the course of the successive stages of the life cycle (eggs, larvae, juveniles and adults);

- Homing behaviour: to complete their life cycle and ensure the survival of the population, and not simply of individuals, a sufficient number of reproducers have to return periodically to predetermined places in their habitat. The return to fry ponds which are stable in time and space thus appears to be a behavioural response to the double imperative of sexual reproduction and life in the aquatic milieu.

These constraints mean that marine species are made up of populations which are distinguished by the succession of patterns of distribution and migration specific to the different stages of the life cycle, their demographic characteristics (average abundance, growth, natural mortality), and their genetic heritage. During the initial stages, the numbers in each population fall dramatically under the effect of environmental processes independent of the abundance of the population (effect of temperature on sexual maturation, gametogenesis and viability of the eggs as in the case of the scallop; drifting into inhospitable hydrodynamic structures where the young stages may either perish or survive but lose any chance of participating in the reproduction of the population, as in the case of several species of fish), and trophic processes depending on the density of the population (predation, famine).

These demographic strategies have very important consequences for the design of high quality research programmes:

- As the regulation processes are concentrated in gametogenesis and the initial stages of the life cycle and are, at least partially, independent of density, it is very probable that for a good many species the biotic capacity of the habitat occupied by the adults is not a limiting factor for the levels of recruitment commonly observed for natural populations; for this reason substantial overpopulation should be compatible with the trophic capacity of the ecosystem. This hypothesis is confirmed by experience (shellfish culture, salmon ranching), as well as by the great amplitude of natural recruitment in certain species;

- From the strictly ecological standpoint, the general nature of homing behaviour and the fact that a majority of marine species reproduce in coastal areas allows us to conclude that sea ranching should be possible for a great variety of species;

- The technical failure of many experimental programmes is probably explained by inadequate taking into account of the particularities of the reproduction strategies of marine populations; massive mortality occurs when the individuals released do not find shelter (lobster, abalone), or the food they need, or fall prey to natural predators (scallop in France, lobster and cod in Norway, etc.);

- The choice of the stock is often critical for the success of restocking and acclimation programmes: as a general rule, the success rate for salmon ranching operations falls when the restocking is made with stocks coming from another ocean or the opposite side of the same ocean. The stocks have to be adapted to the length of the river or stream in order to be able swim up it. Certain poor rates of return are explained by a poor acquisition by cultivated animals of homing behaviour, a component of which is probably genetic. The reproduction strategy of the scallop, for which the success of recruitment depends on synchronisation between gametogenesis and the summer warming cycle in the habitat occupied by each population, could well cause restocking with spat from populations living in very different habitats to be a failure;

- On the other hand, as both juveniles and adults of diverse species may survive in hydrodynamic structures different from those inhabited by the populations to which they belong, and are more tolerant than eggs and larvae, the ecological constraints should generally be less severe for fattening and overpopulation operations not involving the participation of the released individuals in reproduction. Shellfish cultivation in sectors very distant from their area of reproduction is consistent with this hypothesis.

As the ecological constraints change with the objectives, the success of the R&D programmes thus depends on the explicit definition of the objectives, the underlying hypotheses and the experimental protocols. These conditions are difficult to meet in projects where research and development are not clearly distinguished. With advances in knowledge about the reproduction strategies of marine populations and the interest shown by marine ecologists in the full scale ecological experiments constituted by programmes to develop extensive systems, as well as in the possibilities for experimentation on the ecology and physiology of reproduction offered by hatcheries, these aspects are now better taken into account in extensive aquaculture programmes.

Availability of Sites

Competition for sites with already established uses of the coastal strip may be an obstacle to the implantation of new fish farms. Shellfish culture initially developed on the intertidal zone, in bays and basins whose shores were not particularly sought after at the time. In countries where shellfish culture has been implanted for a long time (Spain, France, Netherlands, etc.), a good proportion of the favourable sites have now been colonised. A limited extension of cultivation towards the open sea remains possible thanks to technical innovations (oysters cultivation in deep water, mussel culture on nets). Where it is well established, shellfish culture benefits from the advantage of its anteriority in the allocation of new concessions. This is not the case with new implantations (mussel culture on hurdles) which often comes up against the opposition of local residents. Environmental protection arguments may be put forward with greater force against new forms of cultivation than against traditional uses (fishing, agriculture). These practices reveal the weakness of current methods of allocating the maritime space between competing uses. The fact that several types of extensive cultivation (vagile species and deep sedentary species) make relatively little demand on the maritime space could be an advantage in regions where the coastal strip is very sought after.

Biotic Capacity of the Ecosystems and Water Quality

While, for the reasons explained above, the biotic capacity of ecosystem is generally not an obstacle for the start-up of new extensive systems, this observation is not an absolute rule. The productivity of the ecosystem is not sufficient everywhere to permit growth rates compatible with profitable shellfish farming. Urban and agricultural wastes may, within certain limits, remedy this poverty, just as certain clean-up programmes may reduce the productivity of certain shellfish farming basins.

More critical for the development of extensive aquaculture, and more particularly of shellfish culture, is the quality of the water. Feeding through filtration and bioaccumulation in the food chain make shellfish farming stocks and products very vulnerable to any deterioration in the quality of coastal waters under the effect of urban, agricultural and industrial pollution.

Technical Efficiency and Profitability

To evaluate the technical efficiency of a new system, it is necessary to be able to identify the individuals when they are recaptured. Several marking techniques are used for this purpose: external and internal physical marks, genetic marking (Norwegian PUSH programme for cod), or morphological marking (in Japan, for example, hatchery products may have characteristic traits: special form of the nostrils of sea bream, black pigmentation of the blind face of flounder, etc.). The possibility of changes in the rules of access to the stock and the imposing of a contribution to the financing of alevin rearing may cause fishermen to under-declare the recaptures. With good statistics, the estimation of the impact of rearing programmes on the biomass of a population and the fishing yield does not present any technical difficulties. On the other hand, the segregation of the different causes of mortality and losses (natural mortality, excess mortality of the artificial stock, straying rate, captures by different fleets) may prove difficult.

The existence of a persistent shortage of sea products of high commercial value in the rich countries and the fact that the development of extensive methods is not generally any more difficult for noble species than for others are factors favourable for the development of extensive aquaculture. On the other hand the poor correspondence, frequent in extensive systems, between social and private costs and benefits is a major handicap. It is the primary cause of the apparently poor efficiency of many extensive aquaculture programmes, and of the frequently insuperable difficulties encountered in trying to move from public to private financing. Demonstration of the technical and economic viability of a new system rarely suffices to ensure its take-off.

In order for a programme to be profitable, it suffices in principle for the price of the fish recaptured to exceed the cost of production and recapture of the individuals released. The shortcomings of the exclusion regimes mean that catches by third parties have to be counted as losses (in addition to those due to natural mortality and the straying of individuals), whereas from the technical standpoint they correspond to gains. A not insignificant proportion of both the gains (benefits for the conservation of piscicultural fauna) and the costs (use of the natural biotic capacity, risk of genetic mixing, modification of the landscape, etc.) are social. The financing will be all the more difficult to transfer to the direct private beneficiaries the greater these external effects. This is why a simple financial analysis at the level of the enterprise is not enough to establish the economic and eco-logical value of a new system. A cost/benefit analysis on the scale of the programme as a whole is essential. It will gain from being complemented by a risk analysis at the level of the main actors concerned (direct beneficiaries and other users of the ecosystem).

Ownership of the Stocks

Examination of extensive systems reveals the existence of correlations between the economic and social organisation of the operation and the process of intensification similar to those described for agriculture. Virtually all restocking programmes for the conservation of the fauna are financed by the public sector and benefit the society as a whole without restriction. The same is generally true of breeding young fish to support commercial fisheries in which access remains largely open and free: it is almost always financed by public funds. A partial move towards privatisation is to be seen in sport fishing in which anglers enjoy exclusive rights – collective or individual -, in return for a contribution to the financing of the breeding programmes. Elements of two systems are found in traditional fisheries: even though fishermen may contribute, the cost of the restocking programmes is generally borne by the state, but in common law access is often limited by a system of collective territorial rights. The collective ownership regime existing in Japan provides a good example of this type of regime. Japan is one of the rare countries to have legalised the privileges that coastal fishermen derive from their anteriority of occupation of the coastal strip and exploitation of coast-al stocks. Collectively, fishermen's cooperatives and certain private groupings hold rights of qua-si-ownership over halieutic resources. While they are not permitted to sell their rights, neither may the state withdraw these rights. This formalisation of the common law is not unconnected with the development of extensive aquaculture in this country. While private ownership and exploitation are the exception in the extensive cultivation of vagile species (private farms for exploitation for profit by the owner, or sport fisheries where access is governed by individual rights), they become the rule in shellfish culture, in which the stock is readily appropriable and effectively appropriated. The difference does not end there: shellfish culture is considerably more developed than the ranch-ing of vagile species. While the statistics on this latter are disparate, the figures available show that national productions are often an order of magnitude lower than those of shellfish culture. Lastly, the private ownership of the stock is virtually the rule in intensive fish farms.

Counter-examples of programmes which, despite their technical success, have not led to commer-cial production, or whose the take-off is blocked by institutional constraints, are no less clear. In France, for example, where the extensive culture of the scallop is technically mastered and has been the subject of full-scale experiments financed by the public sector, the transfer of the cost of planting to private operators is progressing only slowly. The trade organisations accept the fact that fishermen should make a contribution, on a voluntary lump-sum basis, to the financing of a collective restocking programme, but they are opposed to the granting of collective concessions on natural beds – though these are on the face of it the most promising -, or to modulation of the individual contributions which would give fishermen rights of recapture proportional to their con-tributions. This would in fact result in the adoption of a system of tradable individual quotas to regulate access to the fishing of natural beds. The start-up of cultivation on individual concessions comes up against the scale of lump-sum charges applied by the administration for the cultivation of oysters and mussels. Would-be fish farmers consider that the scale in force is too high for start-ing up a new system of cultivation where the level of risk is still high.

Thus the privatisation of the stock appears to be a determining factor for the process of intensifi-cation: from fishing to intensive cultivation, individualisation of exploitation (traditional groups, sport fishing associations, commercial fleet owners, private farms), technical intensification (con-servation, fishing, extensive aquaculture, intensive aquaculture), privatisation of the stock (public, collective, and individual ownership), and passage to the monetary sphere (conservation, sport fishing, commercial fishing, aquaculture) all go hand-in-hand.

In primary production where the allocation of the factors is governed by elaborate systems of pri-vate ownership, the rights explicitly define the variables, the holders, the quantities, the offences and the sanctions. They are exclusive in the sense that all the gains and losses inherent in the exer-cise of a right fall to the holders of the right. They are combinable and for this reason tradable. They are technically applicable and effectively applied, since a right which is not applied is not a right.

The mobility of vagile species makes it impossible to simply transpose to extensive cultivation the solution adopted in agriculture and shellfish culture – i.e. the subdivision of the area - for the appro-priation of the stock. However, a better coincidence between the distribution of costs and benefits may be sought by allocating exclusive quantitative rights of alevin rearing and recapture to produc-ers. However the application of this system in cultivation where the same species is fished in the same areas comes up against a problem: the impossibility of distinguishing the cultivated or wild origin of the captures. This uncertainty implies that fishing and cultivation should be subject to the same system of regulation. But it is by no means certain that the increase in production expected from a new system of cultivation will suffice to convince fishermen, who have open and free access to the wild stocks, that it is in their own interest to adopt a system of quantitative individual rights.

The adoption of individual rights also raises the question of the choice of the allocation mech-anism. With the opening up of rural societies to national economies, the systems of customary rights based on social control are gradually falling into disuse. The poor performance of traditional methods of regulation of access in commercial fishing illustrates the ineffectiveness of administra-tive regulation as a way of allocating a factor of production in a commercial activity. The advan-tages of tradable rights (determination of the price of fishing rights by direct negotiation between the parties) comes up against the opposition aroused by the adoption of this system for allocating factors hitherto allocated by non-market mechanisms, and objections of inequality resulting from the people's different capacities for taking advantage of the new opportunities created by the adop-tion of a new system of regulation.

The evolution of ownership regimes and mechanisms for trading rights finally depends on the clarification of national jurisdictions which, at sea as on land, define and guarantee the ownership regimes for natural resources. While the new law of the sea does indeed grant states' rights of quasi-ownership over the resources contained within their exclusive economic zones (EEZs), their authority remains diluted for the stocks whose area of distribution covers the EEZ and the high seas, and is shared for stocks whose area extends over several EEZs.

But for a few exceptions - that of Iceland in particular, which is distinguished by its geographical isolation and the initiatives that this country has taken to adjust the organisation of its fisheries to new conditions -, the present situation in Europe scarcely seems favourable for the extension of extensive aquaculture to vagile species. The great development of continental waters for the moment reduces the prospects for highly migratory species like salmon. In the majority of

European fisheries, access to the resources still remains for the most part open and free. In the EU, the Common Fisheries Policy has so far paid more attention to the maintenance of the status quo and the minimisation of the immediate consequences for fishermen of the recurrent measures taken for the conservation of stocks, than to the clarification of the regime of ownership of the resources and the modernisation of the mechanisms for regulating access.

Rationalisation of Mature Systems

The constraints that extensive systems come up against when they arrive at maturity differ in nature from those which impede their take-off. Because shellfish culture is one of the small number of well-developed systems, our examination of the opportunities and conditions for the rationalisation of mature systems will be based on this mode of production. But the conclusions are transposable to the other systems.

Adjustment of the Stocks to the Biotic Capacity

The hypothesis according to which the trophic capacity of the marine ecosystem should be able to support biomasses greater than those of wild populations does not permit us to predict an increase in the biomass supportable by the various aquacultural ecosystems. The recruitment level of a population and the biotic capacity of an ecosystem to support the juvenile and adult fractions of a population in fact concern different stages of the life cycle and result from distinct processes. Thus the example of shellfish culture shows that the aquacultural potential varies considerably from one coastal ecosystem to another. But whatever the trophic capacity of an ecosystem may be, it is always limited. The finite nature of the natural resource will become increasingly apparent as the biomass in cultivation increases, and will result in a progressive decline in yield growth.

It is thus that in semi-enclosed shellfish farming basins where the water is renewed only slowly, the primary productivity (microphytobenthos and phytoplankton) that the stock feeds on becomes an increasingly limiting factor as the biomass of the stock grows. For access to food for their stocks shellfish farming enterprises find themselves in a situation of competition analogous to that well known in fishing. The continuous movement of the water within a basin prevents the shellfish growers from separately adjusting their stock to the productivity of their concessions. To increase the share of the primary production of the basin that they appropriate they individually have an interest in increasing their stocks, despite the overloading of the basin and the overall decline in yields that this behaviour leads to. Thus, in the Marennes-Oléron basin, the maximum production – slightly over 40,000 tonnes - is obtained for a biomass in the order of 80,000 tonnes. With the Portuguese oyster, the total biomass reached 200 000 tonnes. Exactly the same process was repeated when the basin was replanted with of the Japanese oyster following the disappearance of the Portuguese oyster. The overstocking slows the growth of the oysters - which has the effect of lengthening the fattening cycle (from two to five years) -, and increasing the general mortality rate (multiplying it by two or three). The resulting loss of income only reinforces the overstocking behaviour on the part of oyster growers.

Preservation of Ecosystems

In sectors where shellfish culture is very developed, the farms may modify the circulation of the water (hurdles) and the substrate (sedimentation by the biodeposits - faeces and pseudofaeces).

The productivity of the sites may be affected. In the Bay of Mont St Michel, for example, the deterioration in performance has led the growers to adopt self-imposed limitation of the stocks (spacing of the hurdles).

Epizootic Diseases

The recent history of French shellfish culture illustrates the risk of epizootic diseases connected with the transfer of spat and shellfish between countries and between basins. Between 1967 and 1972, two viruses totally eradicated the Portuguese oyster stocks. Fraudulently introduced, the Japanese oyster is thought to have been the resistant vector of the infectious agent. Through reducing the resistance of the stocks, the overstocking of the basins was probably a facilitating factor, but not the cause of the epizootic diseases. The epidemics appeared, in fact, after peak production levels had been reached in the main basins.

The propagation of diseases has been facilitated by transfers of shellfish between basins. In 1978 and 1982, the flat oyster stocks were hit in their turn by two epidemics whose economic and social consequences were catastrophic. Production fell from over 20,000 tonnes to just a few thousand tonnes. Enterprises had to change species and farming practices and were forced to concentrate. Marketing circuits were disrupted and changed to the benefit of the traditional producers of cupped oysters. For the Brittany region alone, the loss of turnover for the period 1974-82 has been estimated at 1.6 billion current francs.

While the replacement of the Portuguese oyster by the Japanese has made it possible for production to start anew, the present monoculture increases the risk of epizootic diseases, which is all the more serious in that at present there is no substitute species. The difficulty of treating stocks in farms where curative measures are possible only on the spat before its immersion or, with heavy handling costs, on adults in the fattening basins, and the weak advance of mollusc genetics, mean that health strategies depend essentially on prevention. Checks are carried out on imports and also on immersion, transfer between basins, and farming practices and the state of health of the stocks are monitored.

Exotic Species

Despite these controls, exotic species are regularly introduced on the occasion of imports of spat and illegal immersion of shellfish before their commercialisation. The apparitions of non-indigenous are in fact concentrated around the big oyster growing basins: Hydroides ezoensis, Apitasia pulchella, Anomia chinensis, Balanus amphytrite, etc. on the Atlantic coast of France; algae Laminaria Japanica, Undaria pinnatifida and Sargassum muticum in the Thau coastal pond on the Mediterranean coast of France. These imports may also have helped propagate the shelf limpet (Crepidula fornicata) - a competitor of the shellfish stocks - and a type of periwinkle (Ocenebra erinacea) - a predator of the oyster. Similarly, transfers of fish between Sweden and Norway were at the origin of the introduction of a monogene, Gyrodactylus salaris, which has provoked the catastrophic decline of certain wild salmon populations.

As shown by examples of the fortuitous introduction of exotic species, the risk for biodiversity is particularly serious when the plantings concern non-indigenous species. There have been many transplantation experiments in the past: Pacific salmon on the Atlantic coasts of Canada and the USA, Russia, Chile, New Zealand, etc.; Atlantic salmon on the Pacific coast of Canada; Atlantic lobster in the Pacific; king crab from the Sea of Okhotsk in the Barents Sea; South African craw-fish in the Channel; algae (Laminaria, Undaria, etc.) from Japan or Macrocystis from the Pacific coast of America in the Channel, etc. The Japanese oyster (Crassostrea virginica) has colonised practically the entire northern hemisphere. To limit this risk, the International Council for the Exploration of the Sea and the European Consultative Committee for Inland Water Fisheries have published a code and a practical guide applicable to the introduction of non-indigenous species and to the transfer of stocks within the same country. These provisions are not binding however. The example of shellfish culture shows how difficult it is to ensure compliance with the regulations governing transfers once cultivation has reached a high level of development.

Genetic Mixing

Large-scale rearing and restocking programmes which are not rigorously planned may constitute a threat to biodiversity. Through diluting the demographic structures of a species, the invasion of an ecosystem by a common genome may reduce the genetic adaptability of the wild population and cause the disappearance of local populations. The risk depends on the survival and reproduction rates of the foreign individuals, as well as the hybridisation rate and competition of the exotic strains with the wild stocks.

The risk is higher with intensive cultivation, since selection is aimed at artificialising the genetic heritage of the stocks. In the salmon farms of northern Europe, escapes from fish farms in cages already commonly exceed the recruitment of the wild populations, and are capable of exceeding that of the sea ranching programmes: in Scotland and Norway, the escapes from cages are of the same order (2 million individuals) as the number of smolts deliberately released for sea ranching.

In the case of restocking with wild strains already present in the ecosystem colonised, the risk of genetic mixing is theoretically zero. To conserve the biodiversity of the wild populations of anadromous species such as the salmonids, planting may also be limited to one part only of the rivers, the wild populations being conserved in the others. This is what is done in Ireland for trout and salmon. Restocking may also be practised with sterile animals (triploids), but the genetic manipulation of strains intended for human consumption does not have a favourable image in the eyes of the public.

Up to now there is little evidence of mixing between domestic stocks and wild stocks and the modifications observed are mostly minor. In a review of the knowledge available on the genetic impacts of hatchery fish on the wild populations of salmon and steelhead trout in the Pacific, Campton concluded in these terms: "Although widely accepted theoretical and conceptual arguments suggest that the direct genetic effects of hatchery fish on wild populations could be substantial and potentially detrimental, the empirical data supporting those arguments are absent or largely circumstantial." The problem is that for a long time sea ranching programmes were conducted without regard to the identity of the populations, without first establishing the zero state of the genetic heritage of the wild populations, and without genetically marking the individuals released. Since then evaluations of the rate of mixing have been carried out with genetically marked populations, in particular on cod in Norway and trout in Ireland.

In the present state of knowledge, the risk is thus poorly understood. But the fact that public opinion is probably very sensitive to it is in itself a threat to the development of aquaculture. For this reason certain specialists put this risk at the top of the list of the impacts of intensive cultivation on the environment, ahead of organic wastes - faeces and unconsumed food -, antibiotics, medicines and antifouling paints used on the cages. In order to have precise instruments to measure and monitor genetic impacts on natural populations, the Convention on Biodiversity is now preparing an in-depth analysis of the risk created by sea ranching and intensive aquaculture.

Ownership of Natural Resources

When an extensive cultivation system becomes mature, its rationalisation involves the adjustment of the biomass in cultivation to the trophic capacity of the ecosystem. This adjustment has to be made on the scale of the entire ecosystem. This requires that the ownership of the ecosystem - the natural resource - should be clarified. To take an image from agriculture, the breeder should have exclusive rights over his livestock and its pastures in order to manage it properly. The transposition of this model to the sea comes up against to two difficulties, one technical, the other institutional:

- The disparity of scales between the unitary resource - the aquacultural ecosystem -, on the one hand, and the fish farms on the other: the problem may be overcome by attributing the function of adjustment of the total biomass to a single authority - public or collective; for this, the functions which are concerned with the ownership of the natural resource ought to be distinguished from those which concern the use of this resource - i.e. the exploitation of defined fractions of the natural resource;

- The disparity of the mechanisms used to allocate the human inputs (capital and labour), on the one hand, and the fractions of the natural resource, on the other. The former are mobilised by the fish farmers. For this they rely on mechanisms mainly of a market nature. Rights of use, for their part, are traditionally allocated by decisions of the authority responsible for development, in more or less close collaboration with the body of fish farmers. Being a matter of allocating a factor of production in a competitive activity, experience shows that these methods, whether they are used separately or jointly, are less efficient for controlling overexploitation than economic mechanisms - taxation or market mechanisms.

The problem of the clarification of exclusivity rights thus arises at two levels: that of the ownership of the natural resource, and that of the right of use. The problem can be analysed in same way for the other uses of the natural resource: fishing and intensive aquaculture. Only the fraction of the ecosystem used changes with different uses.

The need for clarification of the ownership regime is independent of the identity of the holders of the rights, as of the mechanism for allocating the rights of use. If the regulation of uses is in the hands of a public administration, its efficiency will be reduced if the respective responsibilities of public sector and trade structures, of sectoral administrations (fishing, aquaculture, environment), and different levels of the politico-administrative structure (from local to international), are imprecise. Seen from this angle, the concept of ownership no longer appears as an absolute right over things, but as an assortment of exclusive rights formalised to exercise different controls, allocated according to the respective capacities of the holders (from individual to international) to exercise controls pertinent to the appropriate scales.

While the adoption of the EEZ constitutes a decisive stage in the adjustment of the institutions to new conditions, the new law of the sea remains insufficient to ensure the conservation of renewable resources, the development of new aquacultural systems, and the organisation of fishing. The major shortcomings concern:

- Clarification of the exclusivity regimes:

  o National legislations still fail to make a proper distinction between the ownership and the right of use of natural resources;

  o This imprecision has an impact on the exercise of the functions which attach to these responsibilities:

    ◊ The integrated development of resources is difficult to achieve when it is approached from the standpoint of sectors of activity rather than by complexes of resources: the take-off of the ranching of vagile species requires, as we have seen, the integration of the systems for the allocation of the rights of recapture of the cultured stocks and of fishing the wild stocks. Similarly, to be efficient the adjustment of the aquacultural stocks to the capacity of the ecosystem and the conservation of this capacity should be in the hands of the same structures; it is a matter, in fact, of allocating the potential of the natural ecosystem between competing uses;

    ◊ The supervisory administrations have difficulty in reconciling their responsibility for the conservation of natural resources with that of supporting the professions. In the majority of European fisheries policies, for example, the minimisation of the immediate costs for the profession of the recurrent resource management measures often takes precedence over the introduction of the reforms necessary for the rationalisation of the sector and the conservation of halieutic stocks;

  o In many countries, the organisation of fisheries remains centralised at national level – in the case of the European Union it is at Community level, which is larger than the scale of the great majority of halieutic stocks and ecosystems. This centralisation reduces the efficiency of public interventions (subsidiarity principle). It was justified when the prime responsibility of the administrations was to defend the interests of their national fleets, free to operate in the world ocean, but this is no longer the case since the adjustment of the uses to the potentials of the natural resources has become the main consideration, and now that the extension of national jurisdictions has given national authorities the power to exercise their responsibilities on the scale of the unitary resources;

- Modernisation of the mechanisms for allocating rights of use: the origin of the dynamic of overexploitation being economic, economic mechanisms are generally more efficient for rationalising the commercial uses and, as a result, conserving the resources. For the same reasons, the development of aquaculture will be more dynamic if the rights of use of the natural resources are allocated by economic mechanisms.

Exogenous Factors

Deterioration of the Environment

Shellfish culture provides a good example of the conflicts which arise between extensive aquacultural systems and the utilisation of the carrying capacities of the milieu. These conflicts are not symmetrical. While shellfish culture modifies the environment very little, its localisation in coastal zones where the wastes of human activities accumulate, and the fact that the cultivated species feed by filtration, which concentrates any micro-organisms or toxic substances in the products, makes them very sensitive to any deterioration in the cultivation milieu. This vulnerability may be illustrated by the following examples.

The growth of the stocks and their state of health depend on the quality of the shellfish culture ecosystem:

- At the end of the 70s, the use of tributyl tin paints to protect the hulls of ships and boats against biological fouling caused oyster production in the Arcachon basin (France) to fall from 12,000 to 3,000 tonnes; production returned to its previous level after the use of these paints was banned;

- Plankton bloom (Gyrodinium aureolum) periodically causes high mortality in mollusc cultivation (scallop hatcheries and stocks, for example).

Eating shellfish raw increases the health risk. Purification techniques are, in fact, efficient or sure only for microbial contamination, and not for chemical or biotoxin contamination. This health risk is also liable to lastingly affect the quality image of the products:

- Epidemics caused by the consumption of shellfish may be of bacterial origin (cholera occasioned by the consumption of mussels as in Naples in 1973; typhoid fever; dysentery, etc.) or of viral origin;

- Plankton bloom may synthesise neurotoxins capable of causing paralyses that can be serious for man (Protogonyaulax for example) and toxins capable of causing less serious troubles such as diarrhoea (Dynophysis).

What determines the appearance of toxic bloom has not yet been established. These blooms appear naturally, in fact, but the apparent increase in their manifestations and the concentration of occurrences in the plumes of rivers lead to the suspicion that the origin must be partly anthropic.

Demand

There is a considerable deficit of sea products in the majority of European countries, especially quality products. Given the high nutritive value (low calories, low fat and cholesterol contents) and taste quality of fish, the demand for these products in unlikely to fall. The deficit could even increase if European and world halieutic production continues to decline and if the supply from developing countries levels off. Given the predominance of fishing in the total supply and the time necessary for the institutional adjustments required in the organisation of fisheries, this scenario is very plausible. Even if the expansion of existing fish farms and the take-off of new intensive projects reduce the deficit, diversification will probably not go hand in hand with the increase in production, because intensification is generally accompanied by a reduction in the diversity of the species cultivated. The potential interest of extensive aquaculture should be viewed in this light. Technically, the method can be used to produce a great variety of species (shellfish, prawns, turbot, bream, sturgeon, salmonids, etc.), and the difficulties of development are on average no greater for systems producing species of high commercial value than for any other species.

Extensive aquaculture could also benefit from a better image than that of fishing and intensive aquaculture. As demonstrated by the actions of ecological movements against the fishing of culturally sensitive species, the inability to prevent overexploitation and to manage the conflicts within the fishing industry are starting to tarnish the excellent image this activity once enjoyed. The image of intensive aquaculture suffers from the damage it is accused of doing to the environment (deterioration of landscapes, milieus and aquatic populations). Its capacity to produce quality products is also called into question, an opinion that the recent mishaps in agriculture have done nothing to correct. The high consumption of fishmeal, fresh water (for the production of smolts in salmon culture) and energy in intensive aquaculture may also be detrimental to its image. By comparison, see ranching could appear to be a more environmentally-friendly activity, and one capable of producing a more natural type of food (provided that the quality of the environment is preserved at the same time). Another aspect that should not be ignored in countries where leisure activities are becoming increasingly important is that it may also contribute to the development and diversification of recreational activities.

The health risk and the difficulty of controlling the quality of the products have so far constituted a serious obstacle to the development of international trade in shellfish. The harmonisation of national quality control legislation being pursued by the European Commission should reduce this constraint.

There is thus reason to believe that extensive systems will enjoy strong demand and generally be looked more favourably than the other modes of production. But this conclusion does not mean that such systems will be able to live up to this expectation in practice.

Role of the Public Sector

Research

The development and installation of extensive systems gives rise to complex questions that research, if it is efficient, may help to resolve. The public sector has a particular role to play in the production of knowledge about issues whose scale takes them out of the field of interest of private operators:

- Restocking programmes for the conservation of piscicultural fauna and halieutic stocks: this is typically the case with support programmes for salmonid populations;

- Research and expertise to support the rationalisation of shellfish farming systems: conservation of the biotic capacity and of the environmental quality, adjustment of the biomasses in cultivation to the trophic capacity, fight against epizootic diseases, protection of consumers' health;

- Research on the prospects and conditions for the regulation of access to natural marine resources.

Our examination of the conditions required for take-off and the optimisation of extensive systems highlighted the variety of the disciplines potentially useful for the study of these questions:

- ecology: determination of the recruitment and regulation strategies of natural populations, biotic capacity of aquacultural ecosystems, carrying capacity of aquacultural ecosystems, genetic mixing and contamination between aquacultural stocks and wild populations, fight against epizootic diseases, determination of the causes of plankton bloom, etc.;

- Zootechnics: artificial reproduction, pathology and prophylaxis, genetics, etc. of the species cultivated;

- Economics: cost/benefit analyses of extensive systems, economics of aquaculture and the environment;

- Law of the sea and of the environment: regulation of access to natural marine resources;

- Halieutics – in the agronomic sense of the term -: process of domestication in the marine domain.

It revealed above all that progress in extensive aquaculture depends on the capacity to provide overall responses to complex and coherent sets of questions which differ according to the nature and state of advance of the production systems considered. The lack of coverage and coherence of the research programmes may be considered to be the main cause of past failures. For a long time, targeted research was concerned above all with small-scale questions (such as artificial reproduction) and tended to pay less attention to larger questions (such as the regulation of the natural populations in the aquatic milieu in the biological sciences, or the process of technical intensification, the regulation of access to resources and the role of the ownership in the development and use of renewable marine resources in the human sciences).

Significant efficiency gains may thus be expected from adopting a multidisciplinary approach including the following elements:

- Explicit definition of the objectives: restocking, acclimation, fattening with overpopulation;

- Analysis of culture systems to identify and rank the conditions required for achieving the objectives of development and organisation;

- Identification, conceptualisation and articulation of the pertinent research topics;

- Formulation of hypotheses and of experimental protocols for testing them;

- Execution of the investigations;

- Analysis and integration of the results.

Two other causes of inefficiency frequently cited in the literature concern the lack of rigour in the experimental protocols and the confusion between research and development. These two shortcomings are not independent. While the association of development projects and experimental programmes makes it easier to obtain research funding, this type of funding often hampers the establishment of rigorous experimental protocols and leads to the abandonment of the investigations when the results do not confirm the initial hypotheses.

Research on extensive systems may contribute to the advancement of knowledge on questions of great scientific interest: regulation of marine populations, domestication and technical intensification of aquatic resources, the role of institutions in the exploitation and conservation of these resources. Because they permit full-scale experimentation on the ecology of marine populations, major extensive aquaculture programmes, like those implemented in France on scallops or in Norway on cod, are likely to be of interest to basic research. This interest may facilitate the mobilisation of theoretical and methodological competencies necessary for the study of new questions concerning the exploitation of the living resources of the sea.

Control of the Quality of the Environment and of Aquacultural Stocks and Products

The scale of the controls necessary and concern for the protection of consumers' health justify the intervention of the public sector in three other fields: conservation of environmental quality, preservation of the health of the stocks, and quality control of the products. These three fields are connected with one another. The fact that the state of health of the stocks and the wholesomeness of the products depend upon the quality of the environment explains why the European countries concerned themselves very early – right from the beginning of the century in some cases – with setting up monitoring systems for milieus, stocks and products of cultivation. In 1991, a European directive harmonised the legislation of the EU member states in this field. A system of four levels of wholesomeness was instituted. It has three objectives: conservation of the cultivation milieu, prevention of epizootic diseases, and consumer protection.

Table: New classification of oyster growing areas applied since 1 January 1996 in the European Union.

Areas	Upper limit	Exploitation	
		Natural beds	Cultures
Fit for consumption	300 faecal coliforms/100 g of flesh	authorised (direct consumption)	authorised (direct consumption)
Exploitable	6 000 faecal coliforms/100 g of flesh	authorised (relaying or purification)	authorised (relaying or purification)
Exploitable	60 000 faecal coliforms/100 g of flesh + heavy metals + pesticides	authorised (relaying - > 2 months - or purification)	prohibited (except in the case of derogation)
Not fit for human consumption		prohibited	prohibited

Being reactive, these measures do not address the causes of the degradation of the quality of the coastal milieu or inland seas, only the effects. The passage to preventive requires the adjustment of the institutions.

Adjustment of the Institutions to New Conditions

In the maritime field, technical intensification and the rationalisation of production are hampered by the regimes regulating access to the natural resources. Recent historical studies in the fields of economics and sociology stress role of institutions, and in particular of ownership regimes, in economic growth. For some commentators they are even considered to be fundamental: ' ... innovation, economies of scale, education, capital accumulation, etc..., are not causes of growth, ... Growth will simply not occur unless the existing economic organisation is efficient. Individuals must be lured by incentives to undertake the socially desirable activities. Some mechanisms must be devised to bring social and private rates of return into closer parity ... A discrepancy between private and social benefits or costs means that some third party or parties, without their consent, will receive some of the benefits or incur some of the costs. Such a difference occurs whenever property rights are poorly defined, or are not enforced' . This theory is particularly relevant in the maritime world where, as compared with terrestrial regimes, the institutions governing access to natural resources still have difficulty, despite recent advances in the law of the sea, in adapting to the new situation of the scarcity of natural resources.

At both national and international level, collective initiatives to regulate access are bound to remain uncertain so long as they are not formalised. The problem is that collective decisions do not commit third parties. This is why, over the centuries, when it has been a matter of assuring regalian functions (army, justice, police and money) on which the regulation of access depends, public solutions have always ended up by taking precedence over individual or collective initiatives. To limit theft and freeloading, the application of a common regime has always turned out less costly in the end and public solutions have prevailed.

But the expectation of considerable social benefits does not mean that the institutions necessary for their concretisation emerge spontaneously. Institutional innovations differ in this respect from technical innovations. While the latter may cost a lot, their exploitation is often within the reach of private initiatives, and their benefits go to individuals who are prepared to take the risk. Institutional reforms, for their part, come up against the opposition of vested interests, the inertia of the bureaucracies, and people's apprehensions about the social consequences of change. These constraints have a high immediate cost. But above all, the potential benefits fall to society as a whole, and not to the politicians or administrations who would be prepared to promote them. Under these circumstances advocates of reform are thin on the ground. While the solutions adopted by societies for their organisation may be rational, they do not necessarily seize the opportunities presented to them. The great geographical and sectoral disparities in economic development are indicative of the non deterministic nature of the evolution of societies. Because the adjustment of institutions is part of the political process and depends on public initiatives whose realisation is not a foregone conclusion, the public authorities should direct their efforts at the adjustment of the institutions to new conditions, rather than the support of production and the defence of professional interests.

Intensive Aquaculture

In a semi-intensive system, the production of the pond is increased beyond the level of extensive aquaculture by adding supplementary feed, usually in the form of dry pellets, to integrate the feed naturally available in the pond, allowing for higher stocking density and production per hectare.

In intensive farming, the fish are kept at too high a stocking density to obtain significant amounts of feed from their environment. Instead the fish are dependent on the feed provided and water must be replenished at a high rate to maintain oxygen levels and remove waste. The levels of feed inputs and management of the water affect the stocking density of the fish that can be supported.

In the EU, 80% of farmed fish production comprises the following four species that are predominantly intensively farmed: rainbow trout (reared both in freshwater and at sea) and marine fishes Atlantic salmon , gilthead seabeam and European seabass. In Norway, Atlantic salmon and rainbow trout represent 98% of farmed fish production. Marine fish and rainbow trout are dependent on fish meal and/or fish oil feed. Together they form a relatively small proportion of farmed fish production tonnage, but consume a large part of the global fish oil and fish meal consumed by aquaculture. The intensive farming of these fish, whereby large numbers of them are confined in a small area, causes a range of serious welfare problems.

Pangasius catfish is farmed very intensively in Vietnam, where half of production is for export, the EU and US being leading importers. Vietnamese production of pangasius has increased 10-fold in the last decade to 1.1 million tonnes in 2010 (78% of global farmed pangasius production), of which 0.66 million tonnes was exported. Extremely high stocking densities are made possible by a high rate of water exchange and by the ability of pangasius to breathe atmospheric oxygen, which makes them able to tolerate low levels of dissolved oxygen and highly polluted water. Small-scale pond polyculture systems are being replaced by intensive monocultures that rear pangasius in ponds (which exchange water with nearby river tributaries by tidal exchange and pumping) and in net cages and net pens sited on major river tributaries of the Mekong River delta. Intensive monoculture ponds and net pens are usually stocked at 40-60 fish per square metre, ponds sometimes even higher. Stocking densities for net cages are typically 100-150 fish per square metre.

According to one study, in Vietnam, pangasius are grown in deep ponds (4 m^2 deep) to a market size of 1 kg at stocking densities of 44 fish per m^2. Death rates for these fish are reportedly 20-25%. According to the same study, in Bangladesh, stocking densities for this species are much lower at 6 fish per m2 in 1m deep ponds, and death rates are significantly lower at 10%. Whereas in Vietnam pangasius are farmed in monoculture, the vast majority of farms in Bangladesh also stock around 10–20% filter-feeding carps which eat up algal blooms produced in the fertile ponds (made so by the feed inputs). These carp require dissolved oxygen and are far less tolerant of poor water quality than pangasius, indicating that water quality is maintained at relatively high levels. According to the authors of this study, the better levels of fish welfare (with respect to stocking density, water quality and mortality rates) in Bangladesh also resulted in less use of antibiotics. Both farm-made and commercial feeds, containing wild fish, are widely used in Bangladesh and Vietnam.

Grass carp intensively reared in cages are stocked with other carp (usually as the major species) and the cages are usually about 60 m², with a depth of 2-2.5 m. The fish are fed with aquatic weeds/terrestrial grasses and pelleted or other commercial feeds. Grass carp are stocked at 10 to 20 fish per cubic metre. In addition Wuchang bream are also stocked at 35 to 50 fish per cubic metre. A small quantity of silver and bighead carp are also stocked (1% of the total fish) as 'cage cleaners' (to eat and control algae).

Intensive Sea Farming

Sea cages hold fish captive in a large pocket-shaped net anchored to the bottom and maintained on the surface by a rectangular or circular floating framework. They are widely used for rearing finfish, such as salmon, sea bass and sea bream, and to a lesser extent trout, in coastal and open waters, in areas sheltered from excessive wave action, with sufficiently deep water and relatively low current speeds. Several cages are typically grouped together in rafts, often housing moorings and walkways for boat access, feed storage and feeding equipment. As the water flows freely to the cages, the openness of the system makes it vulnerable to external influences (i.e. pollution events or physical impact) as well as exposing the adjacent environment to the stock, and the fish farm effluents.

Recirculation systems on land can also be used for the farming of marine species.

Fish Farms

A fish farm is where fish are raised for food. This area of agriculture is known as aquaculture. The fish are raised in tanks or in ponds and sold to people or companies.

Some fish that are raised on fish farms are:

- Tilapia
- Cod
- Catfish
- Salmon
- Carp

There are different types of fish farms that utilize different aquiculture methods.

The first method is the cage system which use cages that are placed in lakes, ponds and oceans that contain the fish. This method is also widely referred to as off-shore cultivation. Fish are kept in the cage like structures and are "artificially fed" and harvested. The fish farming cage method has made numerous technological advances over the years, especially with reducing diseases and environmental concerns. However, the number one concern of the cage method is fish escaping and being loose among the wild fish population.

The second method is irrigation ditch or pond systems for raising fish. This basic requirement for this method is to have a ditch or a pond that holds water. This is a unique system because at a small level, fish are artificially fed and the waste produced from the fish is then used to fertilize farmers' fields. On a larger scale, mostly in ponds, the pond is self-sustaining as it grows plants and algae for fish food.

The third method of fish farming is called composite fish culture which is a type of fish farming that allows both local fish species and imported fish species to coexist in the same pond. The number of species depends, but it is sometimes upwards of six fish species in a single pond. The fish species are always carefully chosen to ensure that species can coexist and reduce competition for food.

The fourth method of fish farming is called integrated recycling systems which is considered the largest scale method of "pure" fish farming. This approach uses large plastic tanks that are placed inside a greenhouse. There are hydroponic beds that are placed near the plastic tanks. The water in the plastic tanks is circulated to the hydroponic beds, where the fish feed waste goes to provide nutrients to the plant crops that are grown in the hydroponic beds. The majority of types of plants that are grown in the hydroponic beds are herbs such as parsley and basil.

The last type of fish farming method is called classic fry farming this method is also known as "flow through system". This is when sport fish species are raised from eggs and are put in streams and released.

Cage System

The right choice of site contributes significantly in the success of cage farm. Site selection is vitally important since it can greatly influence economic viability by determining capital outlay, by affecting running costs, rate of production and mortality factors.

- Site selection is a key factor in any aquaculture operation, affecting both success and sustainability.

- Circular cages of different diameter ranging from 2 m to 15 m, designed for the culture of fishes such as milkfish, mullet, cobia, pompano, sea bass, pearl spot, shellfishes such as shrimps, crabs and lobsters were experimented and demonstrated successfully in India by Central Marine Fisheries Research Institute (CMFRI).

- Stocking of right sized fish juveniles in adequate stocking density is another factor which determines the success of farming. The stocking density and size of stocked fishes varies with different species.

- Proper feeding of quality feeds, periodic monitoring and cleaning of cages contributes immensely to the success of cage farming.

- With proper management of cage erected at an ideal location can yield a production of 20-40kg/m^3 with various species of fishes.

Cage aquaculture involves the growing of fishes in existing water resources while being enclosed in a net cage which allows free flow of water. It is an aquaculture production system made of a floating frame, net materials and mooring system (with rope, buoy, anchor etc.) with a round or square shape floating net to hold and culture large number of fishes and can be installed in reservoir, river, lake or sea. A catwalk and handrail is built around a battery of floating cages. There are 4 types of fish-rearing cages namely: i) Fixed cages, ii) Floating cages, iii) Submerged cages and iv) Submersible cages. Economically speaking, cage culture is a low impact farming practice with high returns and least carbon emission activity. Farming of fish in an existing water body removes one of the biggest constraints of fish farming on land, i.e., the need for a constant flow of clean, oxygenated water. Cage farms are positioned in such way to utilize natural currents, which provide the fish with oxygen and other appropriate natural conditions.

Site Selection

Different criteria must be addressed before site selection for cage culture.The physico-chemical parameters like temperature, salinity, oxygen, waveaction, pollution, algal blooms, water exchange, etc. that determine whether a species can thrive in an environment. Other criteria which must be considered for site selection are weather conditions, shelter, depth, substrate, etc. Finally legal aspects, access, proximity to hatcheries or fishing harbor, security, economic, social and market considerations etc. are to be taken care.

Cage Size

It is a fact that costs per unit volume decrease with increasing cage size, within the limits of the materials and construction methods used. CMFRI has developed open sea cages of 6 m dia and 15 m dia for grow out fish culture and 2 m dia HDPE cages for seed rearing. Ideal size for grow out cage is 6 m due to its easy maneuvering and reduced labour. For fingerling, 2m cages can be used.

Cage Frames and Nets

Different cage materials can be used for cage farmes. Materials commonly used are High Density Poly Ethylene (HDPE), Galvanised iron (GI) pipes, PVC pipes, etc. HDPE frames are expensive, but long lasting. Cost effective epoxy coated Galvanized Iron (GI) frames are recommended for Small groups and fishermen. GI frames have less life span when compared to HDPE frames.

Nets of varying dimensions and materials were tested for cage culture in India. CMFRI has used braided and twisted HDPE nets for grow out purpose. It can last for two or more seasons. Nylon net can be used economically, but since it is light weight, to hold the shape intact more weight has to be loaded in the ballast pipe. Cost factor has to be taken care while using new netting materials like sapphire or dyneema materials for net cage. The depth of net ranging from 2 to 5 m is ideal. For open sea cage culture, predator net to prevent attack by predatory organisms is essential.

Potential Species and Criteria for Selection of Species for Cage Culture

The selection of species for cage culture should be based on a number of biological criteria such as omnivore or carnivore, hardiness, fast growing, efficient food conversion ability, availability of quality seeds, disease resistance and market demand.

Stocking

Although stocking densities should be determined by species requirements and operational considerations, the influence of stocking densities on growth and production has been determined empirically. The stocking density depends also on the carrying capacity of the cages and the feeding habits of the cultured species. Optimal stocking density varies with species and size of fish.

Feeds and Feed Management

Fresh or frozen trash fish, moist pellet (MP) and floating dry pellets are the commonly utilized feed for growing fish in cages. Feeding in cages is quite easy compared to that in ponds. The ration can be divided into equal portions and supplied at regular intervals. Feeding can be done either by broadcasting or using feeding trays. Feeds must be nutritionally complete and provide the necessary proteins, carbohydrates, fats, vitamins and minerals needed for growth and health. Feeds cannot be allowed to deteriorate during storage.

Harvest

Harvest of fish in cages is less labour intensive when compared to that in ponds. Floating cages can be towed to a convenient place and full or partial harvest can be carried out based on demand. Marketing of fishes in live conditions as a value addition can also be done.

Cage Management

Cage culture management must result in optimizing production at minimum cost. The management should be so efficient that the cultured fish should grow at the expected rate with respect to feeding rate and stocking density, minimize loss due to disease and predators, monitor environmental parameters and maintain efficiency of the technical facilities. Physical maintenance of cage structures is also of vital importance. The net-cages must be routinely inspected. Necessary repairs and adjustments to anchor ropes and net-cages should be carried out without any delay. Monthly exchange of net should also be considered, as this ensures a good water exchange in the net, thereby washing away faeces, uneaten food and to a certain extent reduce the impact of fouling.

Fouling of Cage Net

Fouling of cage nets and other structures has been observed at many instances of cage farming. Nets get covered with biofoulers. Fouling by molluscs, especially edible oyster and sand barnacles have to be checked before its growth advancement. Algal mats and other periphyton can be removed by introduction of omnivorous grazers in cages. A fouled net will be heavier, thereby increasing drag thus resulting loss of nets and fish.

To avoid/ reduce fouling, net should be changed when required, which may vary from 2 to 4 weeks depending on the intensity of fouling. During oyster fouling, net exchange has to be done immediately after the seasonal spat fall. Herbivorous fish such as rabbit fish (Siganus spp.), pearl spot (Etroplus suratensis) and scat (Scatophagus sp.) can be used to control biofoulers, but their application on a large scale needs to be assessed.

Disease monitoring

Monitoring of fish stock health is essential and early indications can often be observed from changes in behavior, especially during feeding.

Advantages and Disadvantages of Cage Culture

Cage culture of fish has advantages and disadvantages that should be considered carefully before cage production becomes the chosen method. A potential fish farmer can produce fish in an existing pond without destroying sport fishing; does not have to invest large amounts of capital for construction or equipment; and can, therefore, try fish culture without unreasonable risks.

Advantages

Cage culture has advantages which include:

- Many types of water resources can be used, including lakes, reservoirs, ponds, strip pits, streams and rivers which could otherwise not be harvested.

- A relatively low initial investment is required in an existing body of water.

- Harvesting is simplified.

- Observation and sampling of fish is simplified.

- Allows the use of the pond for sport fishing or the culture of other species.

- Less manpower requirement.

- Generation of job opportunities for unemployed youth and women.

- Additional income to fishers during closed seasons.

Disadvantages

Cage culture also has some distinct disadvantages. These include:

- Feed must be nutritionally complete and kept fresh.

- Low Dissolved Oxygen Syndrome (LODOS) is an ever present problem and may require mechanical aeration.

- Fouling of net cage.

- The incidence of disease can be high and diseases may spread rapidly.

- Vandalism or poaching is a potential problem.

- Navigation issues.

- Accumulation of unused feed and excreta will lead to water pollution as well as eutrophication.

- Change in water quality parameters.

- Conflicts within the local community.

- Predation by aquatic mammals and birds.

- Escapement.

- Overcrowding of aquatic organisms in cages.

Copper-alloy Nets

Copper alloy netting improves fish production, leading to healthier fish, higher yields, and lower costs. In addition, copper netting:

- Resists storm damage and lasts longer than traditional netting.

- Reduces predator attacks and fish escapes.

- Stays naturally clean, reduces drag and maintains cage volume.

- Decreases impact of pathogens and parasites, as monitored by farmers.

- Supports sustainable fish farming and is 100% recyclable.

- Minimizes maintenance cost and efforts.

Irrigation Ditch or Pond Systems

These use irrigation ditches or farm ponds to raise fish. The basic requirement is to have a ditch or pond that retains water, possibly with an above-ground irrigation system (many irrigation systems use buried pipes with headers). Using this method, one can store one's water allotment in ponds or ditches, usually lined with bentonite clay. In small systems the fish are often fed commercial fish food, and their waste products can help fertilize the fields. In larger ponds, the pond grows water plants and algae as fish food. Some of the most successful ponds grow introduced strains of plants, as well as introduced strains of fish.

Control of water quality is crucial. Fertilizing, clarifying and pH control of the water can increase yields substantially, as long as eutrophication is prevented and oxygen levels stay high. Yields can be low if the fish grow ill from electrolyte stress.

Integrated Recycling Systems

One of the largest problems with freshwater aquaculture is that it can use a million gallons of water per acre (about 1 m³ of water per m²) each year. Extended water purification systems allow for the reuse (recycling) of local water.

The largest-scale pure fish farms use a system derived (admittedly much refined) from the New Alchemy Institute in the 1970s. Basically, large plastic fish tanks are placed in a greenhouse. A hydroponic bed is placed near, above or between them. When tilapia are raised in the tanks, they are able to eat algae, which naturally grows in the tanks when the tanks are properly fertilized.

The tank water is slowly circulated to the hydroponic beds where the tilapia waste feeds a commercial plant crops. Carefully cultured microorganisms in the hydroponic bed convert ammonia to nitrates, and the plants are fertilized by the nitrates and phosphates. Other wastes are strained out by the hydroponic media, which doubles as an aerated pebble-bed filter.

This system, properly tuned, produces more edible protein per unit area than any other. A wide variety of plants can grow well in the hydroponic beds. Most growers concentrate on herbs (e.g. parsley and basil), which command premium prices in small quantities all year long. The most common customers are restaurant wholesalers.

Since the system lives in a greenhouse, it adapts to almost all temperate climates, and may also adapt to tropical climates. The main environmental impact is discharge of water that must be salted to maintain the fishes' electrolyte balance. Current growers use a variety of proprietary tricks to keep fish healthy, reducing their expenses for salt and waste water discharge permits. Some veterinary authorities speculate that ultraviolet ozone disinfectant systems (widely used for ornamental fish) may play a prominent part in keeping the Tilapia healthy with recirculated water.

A number of large, well-capitalized ventures in this area have failed. Managing both the biology and markets is complicated.

Classic Fry Farming

Classic fry farming is also called a "flow through system Trout and other sport fish are often raised from eggs to fry or fingerlings and then trucked to streams and released. Normally, the fry are raised in long, shallow concrete tanks, fed with fresh stream water. The fry receive commercial fish food in pellets. While not as efficient as the New Alchemists' method, it is also far simpler, and has been used for many years to stock streams with sport fish. European eel (Anguilla anguilla) aquaculturalists procure a limited supply of glass eels, juvenile stages of the European eel which swim north from the Sargasso Sea breeding grounds, for their farms. The European eel is threatened with extinction because of the excessive catch of glass eels by Spanish fishermen and overfishing of adult eels in, e.g., the Dutch IJsselmeer, Netherlands. As per 2005, no one has managed to breed the European eel in captivity.

Fishery

A fishery is an organized effort by humans to catch fish or other aquatic species, an activity known as fishing.

Generally, a fishery exists for the purpose of providing human food, although other aims are possible (such as sport or recreational fishing), or obtaining ornamental fish or fish products such as fish oil.

Regardless of purpose, however, the term fishery generally refers to a fishing effort centered on either a particular ecoregion or a particular species or type of fish or aquatic animal, and usually fisheries are differentiated by both criteria.

Examples would be the salmon fishery of Alaska, the cod fishery off the Lofoten islands or the tuna fishery of the Eastern Pacific.

Most fisheries are marine, rather than freshwater; most marine fisheries are based near the coast.

This is not only because harvesting from relatively shallow waters is easier than in the open ocean, but also because fish are much more abundant near the coastal shelf, due to coastal upwelling and the abundance of nutrients available there.

Types of Fishery

There are two main types of fisheries: Inland fisheries and marine fisheries.

Inland or Fresh Water Fisheries

Inland fishery deals with the fishery aspects of waters other than marine water. Potentially, the vast and varied inland fishery resources of India are one of the richest in the world. They pertain to two types of waters, namely, the fresh and the brackish. The former includes the country's great river systems, an extensive network of irrigation canals, reservoirs, lakes, tanks, ponds, etc.

The estuaries, lagoons and mangrove swamps constitute the brackish type of water. In pisciculture (culture fisheries), which generally pertains to small water bodies, the fish seed has to be sown, tended, nursed, reared and finally harvested when grown to table size. In the case of capture fisheries, which pertain to the rivers, estuaries, large reservoirs, as well as big lakes, man has only to reap without having to sow.

	Name		Name
1.	Rohu (Labeo rohita)	5.	Mangur (Clarias batrachus)
2.	Calbasu (L. calbasu)	6.	Singhi (Heteropneustes fossilis)
3.	Catla (Catla catla)	7.	Malli (Wallago attu)
4.	Singhara (Mystus seenghala)	8.	Mirgala (Cirrhinus mrigala)

Types of Breeding

According to the mode of breeding there are two categories, natural breeding and induced breeding.

(i) Natural Breeding (Bundh breeding):

The natural bundhs are special types of ponds where natural water resource conditions are managed for the breeding of culturable fish. These bundhs are constructed in large low-lying areas to accumulate large quantity of rain water. These bundhs are having an outlet for the exit of excess rain water.

(ii) Induced Breeding:

In artificial method of fertilization ova from the females and the sperms from the males are taken out by artificial mechanical process and the eggs are got fertilized by the sperms. Different methods are used for induced breeding. Here induced breeding by hormones method is briefly described. The gonadotropin hormone (FSH and LH) secreted by pituitary gland influences the maturation of gonads and spawning in fishes.

(iii) Composite Fish Farming:

It is found that if few selected species of fish are stocked together in proper proportion in a pond, total production of fish is increased many times. This mixed farming is called composite farming. It has some advantage-compatible species do not harm each other, all available areas are fully utilised, no competition among different species is found and fish may have beneficial effect on each other. Catla catla, Labeo-rohita and Cirrhina mrigala are surface feeder, column feeder and bottom feeder respectively and are used for composite farming.

Marine Fisheries

Marine fishery deals with the fishery aspects of the sea water or ocean.

Fish Diseases caused by Parasites and Pathogens

1. Bacterial Diseases: Two bacterial diseases are very important.

    (i) Abdominal dropsy of Carps is caused by Aeromonas punctuata.

    (ii) Furunculoris of Salmons and trout's is caused by Aeromonas salminicida.

2. Viral Diseases: Economically most important is the viral haemorrhage septicaemia (VHS) of rainbow trouts.

3. Protozoan Diseases: Main protozoan diseases are caused by Costia, Myxobolus and Trypanosoma.

4. Fungal Diseases: The gill rot (branchyomyces) of carps involves the attack of Saprolegnia on the gills of carps.

5. Worm Diseases: Worms of four groups are parasites on fish. The flatworms (trematodes), tapeworms (cestodes), round worms (nematodes) and thorny-headed worms (acanthocephalans).

6. Common Ectoparasites: Two ectoparasites of fish are most important, the fish lice (Argulus, Lernaea and Ergasilus) and the fish leech (Piscicold). Both parasites weaken fish by feeding on its blood.

Fish Farming Issues

Environmental Problems with Fish Farming

Fish farming is a way to create a much larger amount of fish much more quickly, cheaply and efficiently than with wild caught fish. Unfortunately, when something seems to good to be true, it very often is.

Pollution

This density of fish creates problems like disease and pollution. The biggest source of pollution is the accumulation of fish waste and uneaten food beneath the sea pens which can degrade the quality of the surrounding water.

Like commercial farming operations on land, the density of fish in these pens necessitates certain chemicals to keep animals from getting sick and to keep things clean. The chemicals used in marine aquaculture operations such as medicines like antibiotics and vaccines, disinfectants, and substances used to prevent corrosion of equipment (cages, etc.) can also change the composition of the surrounding aquatic ecosystem.

The amount of pollution from fish farms also depends on how the fish are contained. Open-net, or pen systems, allow for a direct exchange of water, where as "closed contentment" methods have a barrier which filters the water.

Impact of Biodiversity

Another way aquaculture can have a negative impact is by introducing farmed species into the wild and therefore changing the biodiversity of aquatic ecosystems. Even when measures are taken to prevent escapes, predators like birds and sharks, equipment failure, human error, severe weather and other complications mean that escapes of farmed fish are inevitable.

Since farmed fish often have been bred via selective breeding they have a lower genetic variation than wild fish. If they interbreed with the wild fish it can result in a less genetically diverse, and therefore less robust, population.

Another concern is infertile offspring. For example, Atlantic and Pacific salmon belong to different genera and while they can produce offspring, those offspring will be unable to reproduce (like mules). If populations of non-native species become established they compete with native populations for resources such as food and breeding sites.

Since farmed fish are selected and bred for certain genetic criteria like size, quick growth and hardiness, escaped species can become invasive, which has been recognized as one of the main causes of global biodiversity loss. One example of this was the Pacific oyster in the UK, which was introduced into its waters in the 1960s via aquaculture with the idea that it would be a more commercially viable species than the native oyster. Unfortunately, these pacific oysters have spread and created reef formations, forcing out the native oysters and altering the marine environment.

Tilapia Takeover

Another example of the negative effect of fish farming on native fish population and environment is with Tilapia. Tilapia is one of the most common types of farmed fish. Most of our tilapia supply is imported from Latin American and Asia, and in 2015, Americans ate 475 million pounds of tilapia.

Tilapia is a warm-water fish native to Africa, but in the last 60 years the governments of poor tropical countries saw the fish as a solution to control weeds and mosquitoes in lakes and rivers, breeding and releasing tilapia into these areas. They are now seen as a nuisance, as they are one of the "most invasive species known and difficult to get rid of once established," says Aaron McNevin.

In Lake Apoyo in Nicaragua, tilapia escaped from a fish farm and their pollution and feeding reduced the lake's quantity of an aquatic plant called charra, which was an important source of food for the lake's native fish populations. Sixteen years later, the lake's biota are still recovering.

Spread of Disease and Antibiotic use

Because farmed fish are raised on unnatural diets and in small enclosures they often breed disease, which can pass to wild populations. This is becoming an increasingly big problem, as are the solutions often used for these diseases.

Some aquaculture productions rely on prophylactic antibiotics to prevent infections. The use of antibiotics can cause drug resistant bacteria to develop which can spread to wild populations.

Sea Lice

Another common disease is sea lice. Not to be confused with an itchy, stinging rash caused by jellyfish larvae, these sea lice are planktonic marine parasites which feed on many types of fish. There are many species but the common "salmon louse" or lepeophtheirus salmonis, has become a big problem for both wild and farmed salmon populations. About a centimeter in size, the sea lice attach themselves to the outside of a fish and feed on its mucous, blood, and skin.

This can cause serious damage to fins, erosion of skin, constant bleeding, and open wounds at risk of infection. On an adult fish this may be only a nuisance, but for small juvenile salmon (around the size of a finger), sea lice can be fatal.

Before offshore industrial scale fisheries became big business in the 1970s, sea lice were rarely epidemic to fish populations. Of course, when hundreds or even thousands of fish are crowded together in a small area, sea lice, and other diseases can easily spread from fish to fish.

This problem not only impacts food supply and fish industry profits, it is spreading to wild fish populations. One example is the salmon in the Broughton Archipelago, a group of islands 260 miles northwest of Vancouver, British Columbia.

In 2007 the area had 20 active fish farms, which raised between 500,000 and 1.5 million fish each. As juvenile wild salmon swam past these open-net farms on their way down river towards the sea, the sea lice infecting the farmed salmon attached to them. A study done that year found that the number of wild pink salmon were down 80% since 1970 because of sea lice infestations. The study concluded that at this rate the wild salmon in the area would die off in four generations or by 2015. While the conclusions of this study were not without controversy, it did seem that the salmon populations recovered when the farms idled.

Pesticide use

As sea lice became a problem in fisheries around the world, an unfortunately common solution was employed: pesticides. One chemical commonly used was emamectin benzoate, or Slice, which when administered to rats and dogs causes tremors, spinal deterioration and muscle atrophy.

Of course soon the lice became resistant, and Slice only worked in triple doses. Other chemicals like hydrogen peroxide, Salmosan, AlphaMax and Calicide chemicals have been employed instead.

While we know that these chemicals can negatively affect ocean water and plant species, we don't have enough research to know how much of these chemicals are absorbed and retained by the fish and if any of this passes to those who eat the fish.

Fish Farming: Effects on Fish

As you might imagine, most species of fish don't thrive when being raised in extremely cramped pens, fed commercial feed, and treated with pesticides, antibiotics and other chemicals. We now know that these modern practices negatively affect the fish as well as their environment.

Higher Levels of Omega-6

Like all animals, fish are what they eat. The nutrition of our food depends on the nutrition of our food's food. For example, salmon in the wild eat smaller fish, which eat aquatic plants rich in beneficial long chain omega-3 fatty acids DHA and EPA. Farm-raised salmon eat pellets, and as the nutritional quality of pellets varies, so does the nutritional quality of the fish. Often young salmon are fed pellets made from plant and animal sources, and they receive more expensive fish/fish oil enriched pellets later in their lifespan just before harvest.

New commercial fish feeds are more likely to have protein and oils derived from grains and oilseeds (like soybeans and canola) and with less fishmeal and fish oil. The difference in feeds accounts for why one study that measured the omega-3 contentment of fish species from six regions of the US found large variations in the omega-3 content in the five salmon species tested.

In the two farm raised varieties tested the omega-3 ranged from 717 mg to 1,533 mg per 100 grams of fish (which is equal to a 3.5 oz serving). Compared to the wild-caught varieties, these farmed fish tended to have higher levels of omega-3s but only because the farmed salmon have more fat overall, including higher levels of omega-6 polyunsaturated fats and saturated fats.

Feeds from vegetable sources can be more sustainable than fishmeal and fish oil. These are often made from smaller fish, lower on the food chain which are sometimes called reduction, pelagic, or trash fish. To create 1kg (2.2 lbs) of fishmeal it takes 4.5 kg (10 lbs) of smaller fish. In fact, today at least 37% of global seafood is ground up to make feed. In 1948 that number was only 7.7%.

These lower food chain fish are the food for many species of aquatic life, and depleting them may cause serious implications for aquatic ecosystems and other sea animals including birds and mammals.

PCBs and POPs

PCBs (polychlorinated biphenys) are industrial pollutants that find their way into fresh waters and oceans and then are absorbed by aquatic wildlife. PCBs are a type of POP (persistent organic pollutant).

Type 2 diabetes and obesity have been linked to POPs, and certain types increase the risk of stroke in women. PCBs are potential human carcinogens, and known to promote cancer in animals. Other potential health effects include negative effects on the reproductive, nervous, and immune systems plus impaired memory and learning.

One study found that PCB levels in farmed salmon, especially those in Europe were five to ten times higher than PCBs in wild salmon. Follow-up studies haven't confirmed this, and there are now strict rules on contaminant levels in feed ingredients which have lowered PCBs in these fish.

It is best to avoid these chemicals completely, but most PCBs are found in the skin, so if farmed raised fish is the only option available, it is possible to reduce exposure by removing the skin and by avoiding fried fish.

Things to Consider when Selecting Seafood

As if the various concerns associated with fish farming weren't enough, there are other important factors to consider when sourcing any kind of seafood.

Mercury

Mercury toxicity can impact brain development in children and negatively affect cognitive function in adults. Mercury is found in the muscle of the fish. The biomagnification of mercury means that organisms higher on the food chain contain higher levels of the metal.

One way to consume seafood yet reduce mercury exposure is by eating smaller fish lower on the food chain, such as sardines.

The Selenium Myth

Many of us in the real-food community have heard that mercury is only a concern if there is not selenium present in the fish, and since most seafood also has high levels of selenium we shouldn't be concerned about mercury.

Dr. Christopher Shade, recently confirmed in an interview with Chris Kresser that this is not the case. He verified that those who are deficient in selenium will be more susceptible to mercury toxicity; however, having good selenium levels doesn't prevent someone from getting mercury toxicity from seafood. Nor does the selenium in seafood bind the mercury and therefore prevent us from absorbing the toxic metal.

One important way the body rids itself of mercury is via glutathione, the body's self-produced master-antioxidant. It is therefore important to support this pathway by consuming sulfur containing vegetables like onions and brassicas, and good amounts of vitamin C.

Omega-3 Levels

Omega-3s are very important for health, and should be consumed in proper ratio with Omega-6 fatty acids. Statistically, most of us consume too much Omega-6 and not enough Omega-3, which some experts blame as one of the root causes of many modern diseases. Fish are an excellent natural source of Omega-3s, but there is a wide range of levels depending on the fish. When choosing seafood, it helps to know which fish have the highest levels of these beneficial fats.

Sustainability

While farmed fish have obvious drawbacks there are also sustainability concerns about wild caught fish.

One of the major concerns is overfishing which has become a global problem. Obviously it becomes difficult to eat the fish if they don't exist, but fewer populations of certain species can have repercussions for an entire ecosystem.

Another issue is bycatch, which is when non-target animals are caught during fishing. This can include dolphins, sea turtles, birds, sharks, stingrays, and other fish like juvenile fish. The incidence of bycatch can be reduced by the use of selective fishing gear designed to catch only the species selected and implementing measures to return the native species.

Habitat destruction can degrade aquatic ecosystems, as seabed habitats provide shelter and food for a variety of species. One fishing method that is a common culprit is bottom trawling near vulnerable areas like coral reefs or breeding and nursing grounds.

Indoor Fish Farming

An alternative to outdoor open ocean cage aquaculture, is through the use of a recirculating aquaculture system (RAS). A RAS is a series of culture tanks and filters where water is continuously recycled and monitored to keep optimal conditions year round. To prevent the deterioration of water quality, the water is treated mechanically through the removal of particulate matter and biologically through the conversion of harmful accumulated chemicals into nontoxic ones.

Other treatments such as ultraviolet sterilization, ozonation, and oxygen injection are also used to maintain optimal water quality. Through this system, many of the environmental drawbacks of aquaculture are minimized including escaped fish, water usage, and the introduction of pollutants. The practices also increased feed-use efficiency growth by providing optimum water quality.

Top four myths about indoor fish farming are:

- Indoor fish farming hurts the environment

  Environmentalists have good reason to be suspicious of fish farms. Previously, these ocean-bound fish stables have been breeding grounds for diseases that can spread to nearby wild marine life. But, indoor fish farming involving recirculating aquaculture systems (RAS), are actually very sustainable. Indoor fish farming is often considered environmentally friendly because it requires less water and produces less waste. Many systems are rather sophisticated and allow for automatic collection and processing of fish wastes into crop fertilizers. As for the spread of disease, these fish only interact with each other in stabilized water, so disease isn't introduced or passed on.

On land, indoor fisheries can be built closer to cities, so less transportation and fossil fuels are required from the farm to the market. Fish grown in the US for US consumers do not involve shipping frozen fish over long distances. On the other hand, fishing boats and the necessary supply-chain distribution of wild-caught fish has a much higher carbon-footprint.

- Indoor fish farming raises the price for consumption

On the economic side, farm-raised fish are bred to make fish cheaper and more readily available to consumers. Currently, America's aquaculture industry supports only 6% of US food demand, producing primarily oysters, clams, mussels, and some fish.

About 90% of fish consumed by Americans is imported, increasing fish prices and contributing to the country's trade deficit. Additionally, global production of fish is much more inconsistent than indoor raised fish, and is susceptible to weather patterns and biological factors (disease, predators). As a result, prices to distributors fluctuate much more than wild caught fish.

- Fish raised indoors isn't nutritious or tasty

Farm-raised fish have more omegas than fish raised in the wild, due to their higher fat content. Farmed salmon, in many cases, can be just as nutritious as its wild-caught salmon, and can be richer in omega-3s and omega-6s essential fatty acids. As for texture, farmed fish tend to have a little bit more fat in their diet, so they might be a little more tender or softer, compared to a wild-caught fish which may be a little leaner.

According to the Seattle Times, farm-raised salmon had a superior taste profile compared to wild caught salmon. 10 different types of frozen salmon ranging from wild-caught in Alaska to farmed in Norway, were prepared by a head chef and served to tasters from The Washington Post Food section and the D.C. area seafood scene. The panel ranked Costco's frozen farmed salmon the best for taste and texture. The bottom of the list was Costco's frozen wild-caught salmon. Everything else fell between, but farmed-fish took the top five spots, and all the wild-caught rounded out the bottom.

- Farm-raised fish is dangerous to eat because of chemicals and pesticides

As a rule, RAS operations do not normally utilize antibiotics, except under the rare cases where poor management practices causes where disease to be introduced into a specific tank. The chance of this happening is minimized compared to fish raised in outdoor pins, as inputs and outputs are tightly controlled. In any event, were a specific batch of fish to be treated, like in all animal husbandry, the fish would not be sent to market until a certain period had passed and the antibiotics were out of their system.

And unlike wild-caught fish, indoor grown fish do not pose a threat of mercury poisoning, an increasing problem in fishes harvested in some areas of the world. Toxic methylmercury from the coal-burning power plants that rim the Northern Pacific Ocean in the US, Japan, China, and Mexico can enter the bodies of large wild caught fish, while fish hatched and raised indoors do not face this threat.

Recirculating Aquaculture Systems

Recirculation aquaculture is essentially a technology for farming fish or other aquatic organisms by reusing the water in the production. The technology is based on the use of mechanical and biological filters, and the method can in principle be used for any species grown in aquaculture such as fish, shrimps, clams, etc. Recirculation technology is however primarily used in fish farming.

Recirculation is growing rapidly in many areas of the fish farming sector, and systems are deployed in production units that vary from huge plants generating many tonnes of fish per year for consumption to small sophisticated systems used for restocking or to save endangered species.

Recirculation can be carried out at different intensities depending on how much water is recirculated or re-used. Some farms are super intensive farming systems installed inside a closed insulated building using as little as 300 litres of new water, and sometimes even less, per kilo of fish produced per year. Other systems are traditional outdoor farms that have been rebuilt into recirculated systems using around 3 m³ new water per kilo of fish produced per year. A traditional flowthrough system for trout will typically use around 30 m³ per kilo of fish produced per year. As an example, on a fish farm producing 500 tonnes of fish per year, the use of new water in the examples given will be 17 m³ /hour(h), 171 m³ /h and 1 712 m³ /h respectively, which is a huge difference.

The formula has been used for calculating the degree of recirculation at different system intensities and also compared to other ways of measuring the rate of recirculation.

Type of system	Consumption of new water per kg fish produced per year	Consumption of new water per cubic meter per hour	Consumption of new water per day of total system water volume	Degree of recirculation at system vol. recycled one time per hour
Flow-through	30 m³	1 712 m³ /h	1 028 %	0 %
RAS low level	3 m³	171 m³ /h	103 %	95.9 %
RAS intensive	1 m³	57 m³ /h	34 %	98.6 %
RAS super intensive	0.3 m³	17 m³ /h	6 %	99.6 %

Seen from an environmental point of view, the limited amount of water used in recirculation is of course beneficial as water has become a limited resource in many regions. Also, the limited use of water makes it much easier and cheaper to remove the nutrients excreted from the fish as the volume of discharged water is much lower than that discharged from a traditional fish farm. Recirculation aquaculture can therefore be considered a most environmentally friendly way of producing fish at a commercially viable level. The nutrients from the farmed fish can be used as fertilizer on agricultural farming land or as a basis for biogas production.

The term "zero-discharge" is sometimes used in connection to fish farming, and although it is possible to avoid all discharge from the farm of all sludge and water, the waste water treatment of the very last concentrations is most often a costly affair to clean off completely. Thus an application for discharging nutrients and water should always be part of the planning permission application.

Most interesting though, is the fact that the limited use of water gives a huge benefit to the production inside the fish farm. Traditional fish farming is totally dependent on external conditions such as the water temperature of the river, cleanliness of the water, oxygen levels, or weed and leaves drifting downstream and blocking the inlet screens, etc. In a recirculated system these external factors are eliminated either completely or partly, depending on the degree of recirculation and the construction of the plant.

Recirculation enables the fish farmer to completely control all the parameters in the production, and the skills of the farmer to operate the recirculation system itself becomes just as important as his ability to take care of the fish.

Controlling parameters such as water temperature, oxygen levels, or daylight for that matter, gives stable and optimal conditions for the fish, which again gives less stress and better growth. These stable conditions result in a steady and foreseeable growth pattern that enables the farmer to precisely predict when the fish will have reached a certain stage or size. The major advantage of this feature is that a precise production plan can be drawn up and that the exact time the fish will be ready for sale can be predicted. This favours the overall management of the farm and strengthens the ability to market the fish in a competitive way.

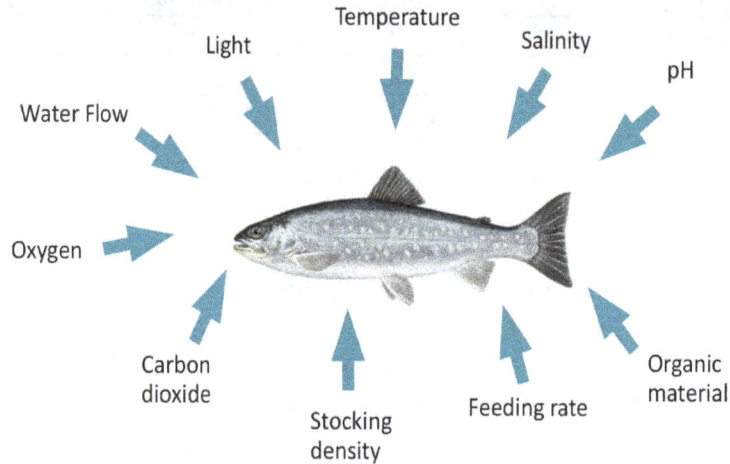

There are many more advantages of using recirculation technology in fish farming. However, one major aspect to be mentioned right away is that of diseases. The impact of pathogens is lowered considerably in a recirculation system as invasive diseases from the outside environment are minimised by the limited use of water. Water for traditional fish farming is taken from a river, a lake or the sea, which naturally increases the risk of dragging in diseases. Due to the limited use of water in recirculation the water is mainly taken from a borehole, drainage system or spring where the risk of diseases is minimal. In fact, many recirculation systems do not have any problems with diseases whatsoever, and the use of medicine is therefore reduced significantly for the benefit of the production and the environment. To reach this level farming practice it is of course extremely important that the fish farmer is very careful about the eggs or fry that he brings on to his farm. Many diseases are carried into systems by taking in infested eggs or fish for stocking. The best way to avoid diseases entering this way, is not to bring in fish from outside, but only bring in eggs as these can be disinfected completely from diseases.

Aquaculture requires knowledge, good husbandry, persistence and sometimes nerves of steel. Shifting from traditional fish farming into recirculation does make many things easier, however at the same time it requires new and greater skills.

Recirculation System

In a recirculation system it is necessary to treat the water continuously to remove the waste products excreted by the fish, and to add oxygen to keep the fish alive and well. A recirculation system is in fact quite simple. From the outlet of the fish tanks the water flows to a mechanical filter and further on to a biological filter before it is aerated and stripped of carbon dioxide and returned to the fish tanks. This is the basic principle of recirculation.

Several other facilities can be added, such as oxygenation with pure oxygen, ultraviolet light or ozone disinfection, automatic pH regulation, heat exchanging, denitrification, etc. depending on the exact requirements.

The basic water treatment system consists of mechanical filtration, biological treatment and aeration/ stripping. Further installations, such as oxygen enrichment or UV disinfection, can be added depending on the requirements.

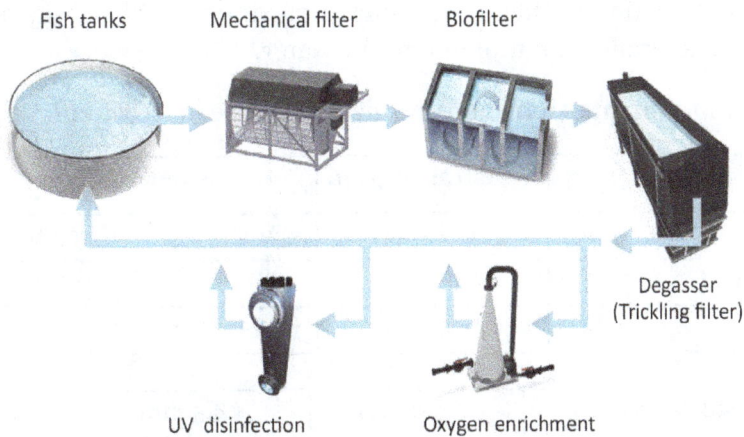

Figure: Principle drawing of a recirculation system

Fish in a fish farm require feeding several times a day. The feed is eaten and digested by the fish and is used in the fish metabolism supplying energy and nourishment for growth and other physiological processes. Oxygen (O_2) enters through the gills, and is needed to produce energy and to break down protein, whereby carbon dioxide (CO_2) and ammonia (NH_3) are produced as waste products. Undigested feed is excreted into the water as faeces, termed suspended solids (SS) and organic matter. Carbon dioxide and ammonia are excreted from the gills into the water. Thus fish consume oxygen and feed, and as a result the water in the system is polluted with faeces, carbon dioxide and ammonia.

Figure: Eating feed and using oxygen results in fish growth and excretion of waste products, such as carbon dioxide, ammonia and faeces

Only dry feed can be recommended for use in a recirculation system. The use of trash fish in any form must be avoided as it will pollute the system heavily and infection with diseases is very likely. The use of dry feed is safe and also has the advantage of being designed to meet the exact biological needs of the fish. Dry feed is delivered in different pellet sizes suitable for any fish stage, and the ingredients in dry fish feed can be combined to develop special feeds for fry, brood stock, grow-out, etc.

In a recirculation system, a high utilization rate of the feed is beneficial as this will minimise the amount of excretion products thus lowering the impact on the water treatment system. In a professionally managed system, all the feed added will be eaten keeping the amount of uneaten feed to a minimum. The feed conversion rate (FCR), describing how many kilos of feed you use for every kilo of fish you produce, is improved, and the farmer gets a higher production yield and a lower impact on the filter system. Uneaten feed is a waste of money and results in an unnecessary load on the filter system. It should be noted that feeds especially suitable for use in recirculation systems are available. The composition of such feeds aims at maximising the uptake of protein in the fish thus minimising the excretion of ammonia into the water.

Table: Ingredients and content of a trout feed suitable for use in a recirculation system.

Pellet size	Fish size, gram	Protein	Fat
3 mm	40 – 125	43%	27 %
4.5 mm	100 – 500	42 %	28 %
6.5 mm	400 – 1200	41 %	29 %

Composition, %	3.0 mm	4.5 mm	6.5 mm
Fishmeal	22	21	20
Fish oil	9	10	10
Rape seed oil	15	15	16
Haemoglobin meal	11	11	11
Peas	5	5	5
Soya	10	11	11
Wheat	12	11	11
Wheat gluten	5	5	5
Other protein concentrates	10	10	10
Vitamins, minerals, etc.	1	1	1

Components in a Recirculation System

Fish Tanks

Table: Different tank designs give different properties and advantages. Rating 1-5, where 5 is the best.

Tank properties	Circular tank	D-ended raceway	Raceway type
Self-cleaning effect	5	4	3
Low residence time of particles	5	4	3
Oxygen control and regulation	5	5	4
Space utilization	2	4	5

The environment in the fish rearing tank must meet the needs of the fish, both in respect of water quality and tank design. Choosing the right tank design, such as size and shape, water depth, self-cleaning ability, etc. can have a considerable impact on the performance of the species reared.

If the fish is bottom dwelling, the need for tank surface area is most important, and the depth of water and the speed of the water current can be lowered (turbot, sole or other flatfish), whereas pelagic living species such as salmonids will benefit from larger water volumes and show improved performance at higher speeds of water.

In a circular tank, or in a square tank with cut corners, the water moves in a circular pattern making the whole water column of the tank move around the centre. The organic particles have a relatively short residence time of a few minutes, depending on tank size, due to this hydraulic pattern that gives a self- cleaning effect. A vertical inlet with horizontal adjustment is an efficient way of controlling the current in such tanks.

In a raceway the hydraulics have no positive effect on the removal of the particles. On the other hand, if a fish tank is stocked efficiently with fish, the self-cleaning effect of the tank design will depend more on the fish activity than on the tank design. The inclination of the tank bottom has little or no influence on the self-cleaning effect, but it will make complete draining easier when the tank is emptied.

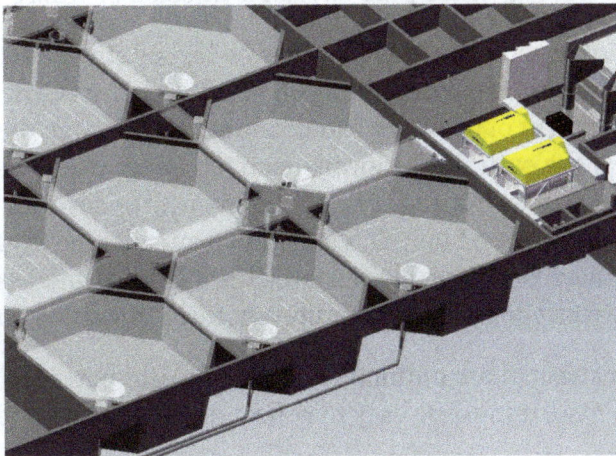

Figure: An example of octagonal tank design in a recirculation system saving space yet achieving the good hydraulic effects of the circular tank

Circular tanks take up more space compared to raceways, which adds to the cost of constructing a building. By cutting off the corners of a square tank an octagonal tank design appears, which will give better space utilization than circular tanks, and at the same time the positive hydraulic effects of the circular tank are achieved. It is important to note that construction of large tanks will always favour the circular tank as this is the strongest design and the cheapest way of making a tank.

A hybrid tank type between the circular tank and the raceway called a "D-ended raceway" also combines the self-cleaning effect of the circular tank with the efficient space utilization of the raceway. However, in practice this type of tank is seldom used, presumably because the installation of the tank requires extra work and new routines in management.

Sufficient oxygen levels for fish welfare are important in fish farming and are usually kept high by increasing the oxygen level in the inlet water to the tank.

Figure: Circular tank, D-ended raceway, and raceway type

Direct injection of pure oxygen in the tank by the use of diffusers can also be used, but the efficiency is lower and more costly.

Control and regulation of oxygen levels in circular tanks or similar is relatively easy because the water column is constantly mixed making the oxygen content almost the same anywhere in the tank. This means that it is quite easy to keep the desired oxygen level in the tank. An oxygen probe placed near the tank outlet will give a good indication of the oxygen available. The time it takes for the probe to register the effect of oxygen being added to a circular tank will be relatively short. The probe must not be placed close to where pure oxygen is injected or where oxygen rich water is fed.

In a raceway, however, the oxygen content will always be higher at the inlet and lower at the outlet, which also gives a different environment depending on where each fish is swimming. The oxygen probe for measuring the oxygen content of the water should always be placed in the area with the lowest oxygen content, which is near the outlet. This downstream oxygen gradient will make the regulation of oxygen more difficult as the time lag from adjusting the oxygen up or down at the inlet to the time this is measured at the outlet can be up to an hour. This situation may cause the oxygen to go up and down all the time instead of fluctuating around the selected level. Installation of modern oxygen control systems using algorithms and time constants will however prevent these unwanted fluctuations.

Tank outlets must be constructed for optimal removal of waste particles, and fitted with screens with suitable mesh sizes. Also, it must be easy to collect dead fish during the daily work routines.

Tanks are often fitted with sensors for water level, oxygen content and temperature for having complete control of the farm. It should also be considered to install diffusers for supplying oxygen directly into each tank in case of an emergency situation.

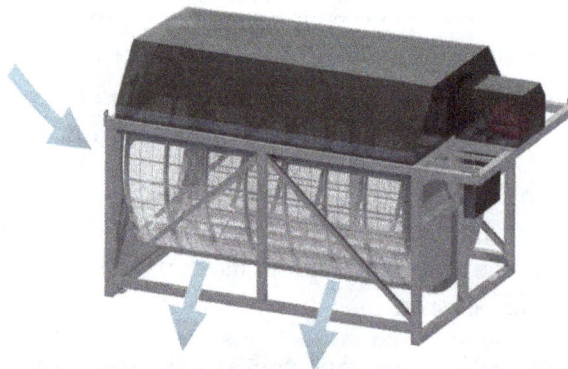

Figure: Drumfilter

Mechanical Filtration

Mechanical filtration of the outlet water from the fish tanks has proven to be the only practical solution for removal of the organic waste products. Today almost all recirculated fish farms filter the outlet water from the tanks in a so called microscreen fitted with a filter cloth of typically 40 to 100 microns. The drumfilter is by far the most commonly used type of microscreen, and the design ensures the gentle removal of particles.

Function of the Drumfilter

1. Water to be filtered enters the drum.

2. The water is filtered through the drum's filter elements. The difference in water level inside/outside the drum is the driving force for the filtration.

3. Solids are trapped on the filter elements and lifted to the backwash area by the rotation of the drum.

4. Water from rinse nozzles is sprayed from the outside of the filter elements. The rejected organic material is washed out of the filter elements into the sludge tray.

5. The sludge flows together with water by gravity out of the filter escaping the fish farm for external waste water treatment.

Microscreen filtration has the following advantages:

- Reduction of the organic load of the biofilter.

- Making the water clearer as organic particles are removed from the water.

- Improving conditions for nitrification as the biofilter does not clog.

- Stabilising effect on the biofiltration processes.

Biological Treatment

Not all the organic matter is removed in the mechanical filter, the finest particles will pass through together with dissolved compounds such as phosphate and nitrogen. Phosphate is an inert substance, with no toxic effect, but nitrogen in the form of free ammonia (NH_3) is toxic, and needs to be transformed in the biofilter to harmless nitrate. The breakdown of organic matter and ammonia is a biological process carried out by bacteria in the biofilter. Heterotrophic bacteria oxidise the organic matter by consuming oxygen and producing carbon dioxide, ammonia and sludge. Nitrifying bacteria convert ammonia into nitrite and finally to nitrate.

The efficiency of biofiltration depends primarily on:

- The water temperature in the system.

- The pH level in the system.

To reach an acceptable nitrification rate, water temperatures should be kept within 10 to 35°C (optimum around 30°C) and pH levels between 7 and 8. The water temperature will most often depend on the species reared, and is as such not adjusted to reach the most optimal nitrification rate, but to give optimal levels for fish growth. Regulation of pH in relation to biofilter efficiency is however important as lower pH level reduces the efficiency of the biofilter. The pH should therefore be kept above 7 in order to reach a high rate of bacterial nitrifying. On the other hand, increasing pH will result in an increasing amount of free ammonia (NH_3), which will enhance the toxic effect. The aim is therefore to find the balance between these two opposite aims of adjusting the pH. A recommended adjustment point is between pH 7.0 and pH 7.5.

Two major factors affect the pH in the water recirculation system:

- The production of CO_2 from the fish and from the biological activity of the biofilter.

- The acid produced from the nitrification process.

Result of nitrification process:

$$NH_4 \text{(ammonium)} + 1.5\,O_2 \longrightarrow NO_2 \text{(nitrite)} + H_2O + 2H^+ + 2e$$
$$\underline{NO_2 \text{(nitrite)} + 0.5\,O_2 \longrightarrow NO_3 \text{(nitrate)} + e}$$
$$NH_4 + 2O_2 \quad \text{«} NO_3 + H_2O + 2H$$

CO_2 is removed by aeration of the water, whereby degassing takes place. This process can be accomplished in several ways.

The nitrifying process produces acid (H+) and the pH level falls. In order to stabilize the pH, a base must be added. For this purpose lime or sodium hydroxide (NaOH) or another base needs to be added to the water.

Fish excretes a mixture of ammonia and ammonium (Total Ammonia Nitrate (TAN)

= ammonium (NH_4^+) + ammonia (NH_3) where ammonia constitutes the main part of the excretion. The amount of ammonia in the water depends however on the pH level as can be seen in figure, which shows the equilibrium between ammonia (NH_3) and ammonium (NH_4^+).

Figure: The equilibrium between ammonia (NH_3) and ammonium (NH_4) at 20°C.
The toxic ammonia is absent at pH below 7, but rises fast as pH is increased

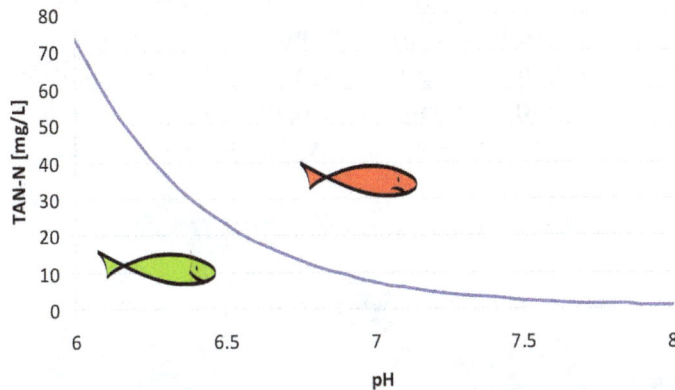

Figure: The relation between measured pH and the amount of TAN available for breakdown
in the biofilter, based upon a toxic ammonia concentration of 0.02 mg/L

In general, ammonia is toxic to fish at levels above 0.02 mg/L. Figure above shows the maximum concentration of TAN to be allowed at different pH levels if a level below 0.02 mg/L of ammonia is to be ensured. The lower pH levels minimises the risk of exceeding this toxic ammonia limit of 0.02 mg/L, but the fish farmer is recommended to reach a level of minimum pH 7 in order to reach a higher biofilter efficiency as explained earlier. Unfortunately, the total concentration of TAN to be allowed is thereby significantly reduced as can be seen in figure. Thus there are two opposite working vectors of the pH that the fish farmer has to take into consideration when tuning his biofilter.

Nitrite (NO_2) is formed at the intermediate step in the nitrification process, and is toxic to fish at levels above 2.0 mg/L. If fish in a recirculation system are gasping for air, although the oxygen concentration is fine, a high nitrite concentration may be the cause. At high concentrations, nitrite is transported over the gills into the fish blood, where it obstructs the oxygen uptake. By adding salt to the water, reaching as little as 0.3 ‰, the uptake of nitrite is inhibited.

Nitrate (NO_3) is the end-product of the nitrification process, and although it is considered harmless, high levels (above 100 mg/L) seem to have a negative impact on growth and feed conversion. If the exchange of new water in the system is kept very low, nitrate will accumulate, and unacceptable levels will be reached. One way to avoid the accumulation is to increase the exchange of new water, whereby the high concentration is diluted to a lower and trouble-free level.

On the other hand, the whole idea of recirculation is saving water, and in some instances water saving is a major goal. Under such circumstances, nitrate concentrations can be reduced by de-nitrification. Under normal conditions, a water consumption of more than 300 litres per kg feed used is sufficient to dilute the nitrate concentration. Using less water than 300 litres per kg feed makes the use of denitrification worth considering.

The most predominant denitrifying bacteria are called Pseudomonas. This is an anaerobic (no oxygen) process reducing nitrate to atmospheric nitrogen. In fact, this process removes nitrogen from the water into the atmosphere, whereby the load of nitrogen into the surrounding environment is reduced. The process requires an organic source (carbon), for example wood alcohol (methanol) that can be added to a denitrification chamber. In practical terms 2.5 kg of methanol is needed for each kg nitrate (NO_3-N) denitrified.

Most often the denitrification chamber is fitted with biofilter media designed with a residence time of 2-4 hours. The flow must be controlled to keep outlet oxygen concentration at app. 1 mg/L. If oxygen is completely depleted extensive production of hydrogen sulphide (H_2S) will take place, which is extremely toxic to fish and also bad smelling (rotten egg). Resulting production of sludge is quite high, and the unit has to be back-washed, typically once a week.

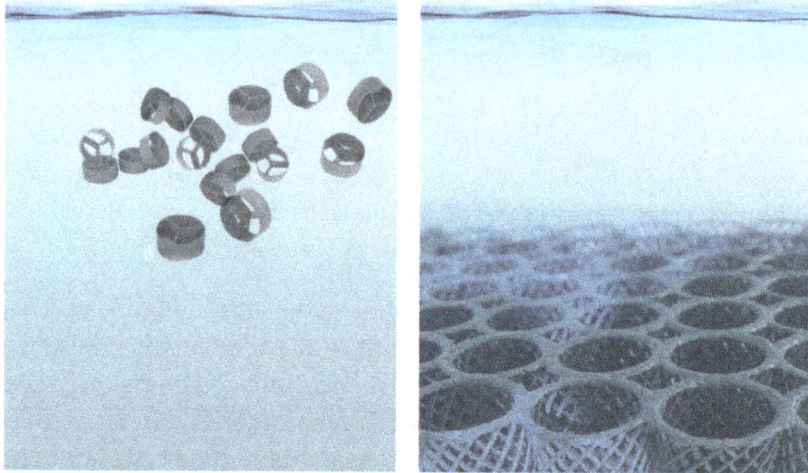

Figure: Moving bed media on left and fixed bed media on right

Biofilters are typically constructed using plastic media giving a high surface area per m3 of biofilter. The bacteria will grow as a thin film on the media thereby occupying an extremely large surface area. The aim of a well-designed biofilter is to reach as high a surface area as possible per m3 without packing the biofilter so tight that it will get clogged with organic matter under operation. It is therefore important to have a high percentage of free space for the water to pass through and to have a good overall flow through the biofilter together with a sufficient back-wash procedure. Such back-wash procedures must be carried out at sufficient intervals once a week or month depending on the load on the filter. Compressed air is used to create turbulence in the filter whereby organic matter is ripped off. The biofilter is shunted while the washing procedure takes place, and the dirty water in the filter is drained off and discharged before the biofilter is connected to the system again.

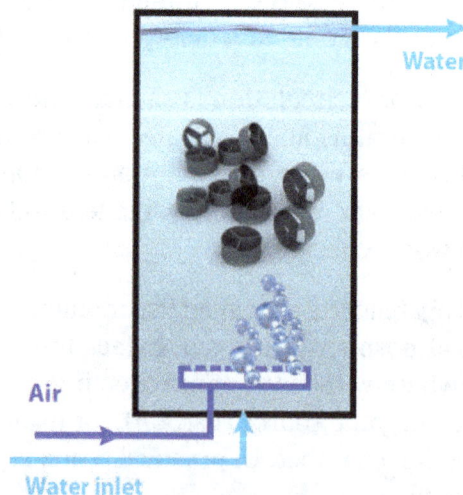

Biofilters used in recirculation systems can be designed as fixed bed filters or moving bed filters. All biofilters used in recirculation today work as submerged units under water. In the fixed bed filter, the plastic media is fixed and not moving. The water runs through the media as a laminar flow to make contact with the bacterial film. In the moving bed filter, the plastic media is moving around in the water inside the biofilter by a current created by pumping in air. Because of the constant movement of the media, moving bed filters can be packed harder than fixed bed filters thus reaching a higher turnover rate per m³ of biofilter. There is however no significant difference in the turnover rate calculated per m² (filter surface area) as the efficiency of the bacterial film in either of the two types of filter is more or less the same. In the fixed bed filter, however, fine organic particles are also removed as these substances adhere to the bacterial film. The fixed bed filter will therefore act also as a fine mechanical filtration unit removing microscopic organic material and leaving the water very clear. The moving bed filter will not have the same effect as the constant turbulence of water will make any adhesion impossible.

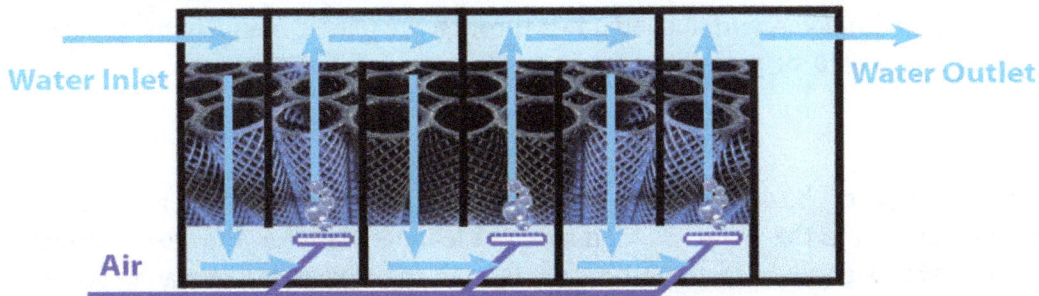

Both filter systems can be used in the same system, or they can be combined; using the moving bed to save space and the fixed bed to benefit from the adhering effect. There are several solutions for the final design of biofilter systems depending on farm size, species to be cultured, sizes of fish, etc.

Degassing, Aeration and Stripping

Before the water runs back to the fish tanks accumulated gases, which are detrimental to the fish, must be removed. This degassing process is carried out by aeration of the water, and the method is often referred to as stripping.

The water contains carbon dioxide (CO_2) from the fish respiration and from the bacteria in the biofilter in the highest concentrations, but free nitrogen (N_2) is also present. Accumulation of carbon dioxide and nitrogen gas levels will have detrimental effects on fish welfare and growth. Under anaerobic conditions hydrogen sulphide (H_2S) can be produced, especially in saltwater systems. This gas is extremely toxic to fish, even in low concentrations, and fish will be killed if the hydrogen sulphide is generated in the system.

Aeration can be accomplished by pumping air into the water whereby the turbulent contact between the air bubbles and the water drives out the gases. This underwater aeration makes it possible to move the water at the same time, for example if an aeration well system is used.

Figure: Trickling filter wrapped in a blue plastic liner to eliminate splashing on the floor. The aeration/stripping process is also called CO_2-stripping. The media in the trickling filter typically consists of the same type of media as used in fixed bed biofilters

The aeration well system is however not as efficient for removing gases as the trickling filter system, also called a degasser. In the trickling system, gases are stripped off by physical contact between the water and plastic media stacked in a column. Water is led to the top of the filter over a distribution plate with holes, and flushed down through the plastic media to maximise turbulence and contact, the so called stripping process.

Oxygenation

The aeration process of the water, which is the same physical process as degassing or stripping, will add some oxygen to the water through simple exchange between the gases in the water and the gases in the air depending on the saturation level of the oxygen in the water. The equilibrium of oxygen in water is 100% saturation. When the water has been through the fish tanks, the oxygen content has been lowered, typically down to 70%, and the content is reduced further in the biofilter. Aeration of this water will typically bring the saturation up to around 90%, in some systems 100% can be reached. Oxygen saturation higher than 100% in the inlet water to the fish tanks is however often preferred in order to have sufficient oxygen available for a high and stable fish growth. Saturation levels above 100% call for a system using pure oxygen.

Figure: Oxygen cone for dissolving pure oxygen at high pressure and a sensor (probe) for measuring the oxygen saturati on of the water

Pure oxygen is often delivered in tanks in the form of liquid oxygen, but can also be produced on the farm in an oxygen generator. There are several ways of making super-saturated water with oxygen contents reaching 200-300 %. Typically high pressure oxygen cone systems or low head oxygen systems, such as oxygen platforms are used. The principle is the same. Water and pure oxygen are mixed under pressure whereby the oxygen is forced into the water. In the oxygen cone the pressure is accomplished with a pump creating a high pressure of typically around 1.4 bar in the cone. Pumping water under pressure into the oxygen cone consumes a lot of electricity. In the oxygen platform the pressure is much lower, typically down to about 0.1 bar, and water is simply pumped through the box mixing water and oxygen. The difference in the two kinds of systems is that the oxygen cone solution uses only a part of the circulating water for oxygen enrichment, whereas the oxygen platform is used for the main recirculation flow often in combination with the overall pumping of water round in the system.

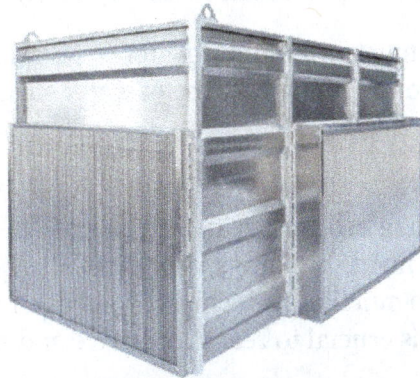

Figure: Oxygen platform for dissolving pure oxygen at low pressure while pumping water around in the farm. The system typically increases the level dissolved oxygen to just above 100% when entering the fish tanks depending on flow rates and farm design

Whatever method is used, the process should be controlled with the help of oxygen measurement. The best way of doing so is to have the oxygen probe measuring after the oxygenation system at normal atmospheric pressure, for example in a measurement chamber delivered by the supplier. This makes the measurement easier than if it was made under pressure, since the probe will need to be wiped clean and calibrated, from time to time.

Ultraviolet Light

UV disinfection works by applying light in wavelengths that destroy DNA in biological organisms. In aquaculture pathogenic bacteria and one-celled organisms are targeted. The treatment has been used for medical purposes for decades and does not impact the fish as UV treatment of the water is applied outside the fish production area. It is important to understand that bacteria grow so rapidly in organic matter that controlling bacterial numbers in traditional fish farms has limited effect. The best control is achieved when effective mechanical filtration is combined with a thorough biofiltration to effectively remove organic matter from the process water, thus making the UV radiation work efficiently.

The UV dose can be expressed in several different units. One of the most widely used is micro Watt-seconds per cm² (μWs/cm²). The efficiency depends on the size and species of the target organisms and the turbidity of the water. In order to control bacteria and viruses the water needs to be treated with roughly 2 000 to 10 000 μWs/cm² to kill 90% of the organisms, fungi will need 10 000 to 100 000 and small parasites 50 000 to 200 000 μWs/cm²

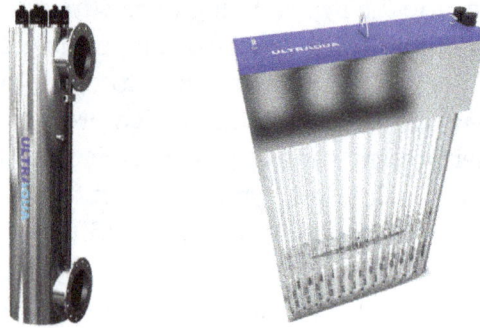

Figure: Closed and open UV treatment systems: For installation in a closed piping system and
in an open channel system respectively

UV lighting used in aquaculture must work under water to give maximum efficiency, lamps fitted outside the water will have little or no effect because of water surface reflection.

Ozone

The use of ozone (O_3) in fish farming has been criticised because the effect of over-dosing can cause severe injury to the fish. In farms inside buildings, ozone can also be harmful to the people working in the area as they may inhale too much ozone. Thus correct dosing and monitoring of the loading together with proper ventilation is crucial to reach a positive and safe result.

Ozone treatment is an efficient way of destroying unwanted organisms by the heavy oxidation of organic matter and biological organisms. In ozone treatment technology micro particles are broken down into molecular structures that will bind together again and form larger particles. By this form of flocculation, microscopic suspended solids too small to be caught can now be removed from the system instead of passing through the different types of filters in the recirculation system. This technology is also referred to as water polishing as it makes the water clearer and free of any suspended solids and possible bacteria adhering to these. This is especially suitable in hatchery and fry systems growing small fish, which are sensitive to micro particles and bacteria in the water.

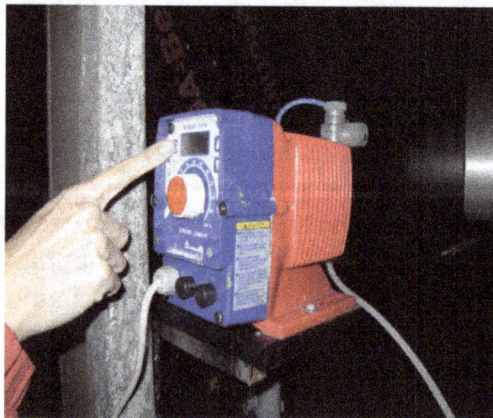

Ozone treatment can also be used when the intake water to a recirculation system needs to be disinfected.

It is worth mentioning that in many cases UV treatment is a good and safe alternative to ozone.

pH Regulation

The nitrifying process in the biofilter produces acid, thus the pH level will drop. In order to keep a stable pH a base must be added to the water. In some systems a lime mixing station is installed dripping limewater into the system and thereby stabilizing pH. An automatic dosage system regulated by a pH-meter with a feedback impulse to a dosage pump is another option. With this system it is preferable to use sodium hydroxide (NaOH) as it is easy to handle and making the system easier to maintain. Sodium hydroxide is a strong alkaline that can severely burn eyes and skin. Safety precautions must be taken, and glasses and gloves must be worn while handling this and other strong acids and bases.

Water Temperature Regulation

Maintaininganoptimalwatertemperatureintheculturesystemismostimportant as the growth rate of the fish is directly related to the water temperature. Using the intake water is a fairly simple way of regulating the temperature from day to day. In an indoor recirculation system the heat will slowly build up in the water, because energy in the form of heat is released from the fish metabolism and the bacterial activity in the biofilter. Heat from friction in the pumps and the use of other installations will also accumulate. High temperatures in the system are therefore often a problem in an intensive recirculation system. By adjusting the amount of cool fresh intake water into the system, the temperature can be regulated in a simple way.

If cooling by the use of intake water is limited a heat pump can be used. The heat pump will utilize the amount of energy normally lost in the discharge water or in the air leaving the farm. The energy is then used for cooling the circulating water inside the farm. A similar way of lowering heating/cooling cost can be achieved by recovering the energy by the use of a heat exchanger. Energy in the discharge water from the farm is transferred to the cold incoming intake water or vice versa. This is done by passing both streams into the heat exchanger where the warm outlet water will lose energy and heat up the cold intake water, without mixing the two streams. Also on the ventilation system a heat exchanger for air can be mounted utilizing energy from the out-going air and transferring it to the in-going air, thereby reducing the need for heating significantly.

In cold climates heating of the water can be necessary. The heat can come from any source like an oil or gas boiler and is, independent of energy source, connected to a heat exchanger to heat the recirculated water. Heat pumps are an environmental friendly heating solution, and can utilize energy for heating from the ocean, a river, a well or the air. It can even be used to transfer the energy from one recirculation system to another, and thereby heat one system and cool another. Usually it utilizes energy from e.g. the ocean using a titanium heat exchanger, moves the energy to the recirculation that is calling for heating and releases the heat through another heat exchanger.

Different types of pumps are used for circulating the process water in the system. Pumping normally requires a substantial amount of electricity, and low lifting heights and efficient and correctly installed pumps are important to keep running costs at a minimum.

The lifting of water should preferably occur only once in the system, whereby the water runs by gravity all the way through the system back to the pump sump. Pumps are most often positioned in front of the biofilter system and the degasser as the water preparation process starts here. In any case, pumps should be placed after the mechanical filtration to avoid breaking the solids coming from the fish tanks.

Calculation of the total lifting height for pumping is the sum of the actual lifting height and the pressure losses in pipe runs, pipe bends and other fittings. This is also called the dynamic head. If water is pumped through a submerged biofilter before falling down through the degasser, a counter pressure from the biofilter will also have to be accounted for.

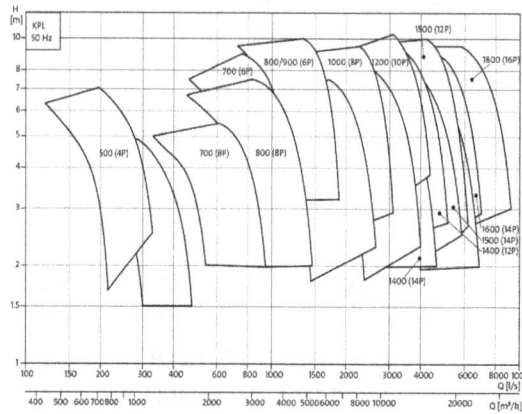

Figure: Lifting pumps type KPL for efficient lifting of large amounts of water

In figure lifting pumps are often used for pumping the main flow in the recirculation system. Correct selection of pump is important to keep the running costs down. Frequency control is an option to regulate the exact flow needed depending on the fish production. H is the lifting height and Q is the volume of water lifted.

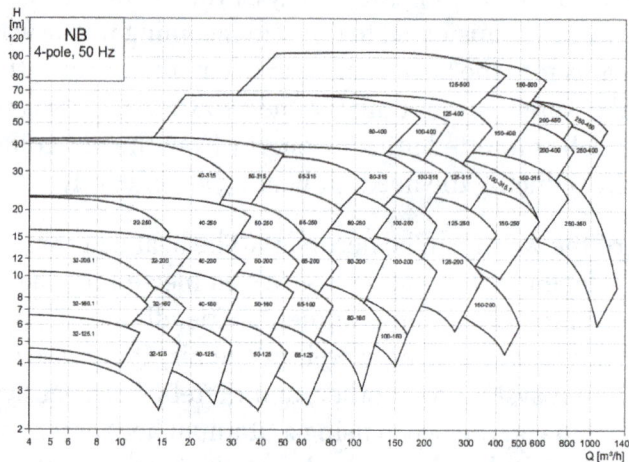

Figure: Centrifugal pumps type NB for pumping water when high pressure or high lifting heights are needed

In figure the range of centrifugal pumps is wide, so these pumps are also efficiently used for pumping at lower lifting heights. Centrifugal pumps are often used in recirculation systems for pumping secondary flows as for example flows through UV systems or for reaching high pressure in oxygen cones. H is the lifting height and Q is the volume of water lifted

The total lifting height in most intensive recirculation systems today is around 2-3 metres, which makes the use of low pressure pumps most efficient for pumping the main flow around. However, the process of dissolving pure oxygen into the process water requires centrifugal pumps as these pumps are able to create the required high pressure in the cone. In some systems, where the lifting height for the main flow is very low, the water is driven without the use of pumps by blowing air into aeration wells. In these systems the degassing and the movement of water are accomplished in one process, which makes low lifting heights possible. The efficiency of degassing and moving of water is however not necessarily better than that of pumping water up over the degasser, because the efficiency of aeration wells in terms of using energy and the degassing efficiency is lower than using lifting pumps and stripping or trickling the water.

Monitoring, Control and Alarms

Intensive fish farming requires close monitoring and control of the production in order to maintain optimal conditions for the fish at all times. Technical failures can easily result in substantial losses, and alarms are vital installations for securing the operation.

In many modern farms, a central control system can monitor and control oxygen levels, temperature, pH, water levels and motor functions. If any of the parameters moves out of the preset hysteresis values, a start/stop process will try to solve the problem. If the problem is not solved automatically, an alarm will start. Automatic feeding can also be an integrated part of the central control system. This allows the timing of the feeding to be coordinated precisely with a higher dosage of oxygen as the oxygen consumption rises during feeding. In less sophisticated systems, the monitoring and control is not fully automatic, and personnel will have to make several manual adjustments.

Figure: An oxygen probe (Oxyguard) is calibrated in the air before being lowered into the water for on-line measurement of the oxygen content of the water

Whatever the case, no system will work without the surveillance of the personnel working on the farm. The control system must therefore be fitted with an alarm system, which will call the personnel if any major failures are about to occur. A reaction time of less than 20 minutes is recommended, even in situations where automatic back-up systems are installed.

Emergency System

The use of pure oxygen as a back-up is the number one safety precaution. The installation is simple, and consists of a holding tank for pure oxygen and a distribution system with diffusers fitted in all tanks. If the electricity supply fails a magnetic valve pulls back and pressurized oxygen flows to each tank keeping the fish alive. The flow sent to the diffusers should be adjusted beforehand, so that the oxygen in the storage tank in an emergency situation lasts long enough for the failure to be corrected in time.

Figure: Oxygen tank and emergency electrical generator

To back up the electrical supply, a fuel driven electrical generator is necessary. It is very important to get the main pumps in operation as fast as possible, because ammonia excreted from the fish will build up to toxic levels when the water is not circulating over the biofilter. It is therefore important to get the water flow up and running within an hour or so.

Intake Water

Water used for recirculation should preferably come from a disease-free source or be sterilised before going into the system. In most cases it is better to use water from a borehole, a well, or something similar than to use water coming directly from a river, lake or the sea. If a treatment system for intake water needs to be installed, it will typically consist of a sand filter for microfiltration and a UV or ozone system for disinfection.

Fish Species in Recirculation

A recirculation system is a costly affair to build and to operate. There is competition on markets for fish and production must be efficient in order to make a profit. Selecting the right species to produce and constructing a well-functioning system are therefore of high importance. Essentially, the aim is to sell the fish at a high price and at the same time keep the production cost at the lowest possible level.

Water temperature is one of the most important parameters when looking at the feasibility of fish farming, because fish are cold blooded animals. This means that fish have the same body temperature as the temperature of the surrounding water. Fish cannot regulate their body temperature like pigs, cows or other farmed animals. Fish simply do not grow well when the water is cold; the warmer the water, the better the growth. Different species have different growth rates depending on the water temperature, and fish also have upper and lower lethal temperature limits. The farmer must be sure to keep his stock within these limits or the fish will die.

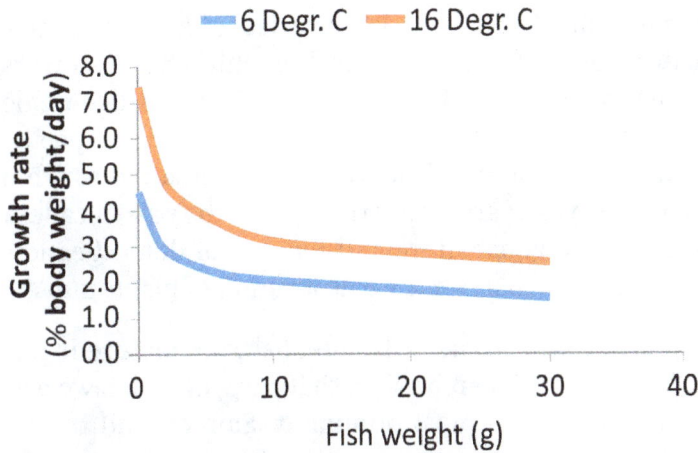

Figure: Growth rate of rainbow trout at 6 degrees and at 16 degrees Celsius as function of fish size

Another issue affecting the feasibility of fish farming is the size of the fish grown in the farm. At any given temperature, small fish have a higher growth rate than large fish. This means that small fish are able to gain more weight over the same period of time than large fish.

Small fish also convert fish feed at a better rate than large fish - see figure. Growing faster and utilising feed more efficiently will of course have a positive influence on the production costs as these are lowered when calculated per kilo of fish produced. However, the production of small fish is just one step in the whole production process through to marketable fish. Naturally, not all fish produced in fish farming can be small fish, and the potential for growing small fish is therefore limited. Nevertheless, when discussing what kind of fish to produce in recirculation systems, the answer, first and foremost, will be small fish. It simply makes sense to invest money in fry production, because you get more out of your investment when farming small fish.

The cost of reaching and maintaining the optimal water temperature all year round in a recirculation facility is money well spent. Keeping fish at optimal rearing conditions will give a much higher growth rate in comparison to the often sub-optimal conditions in the wild. Also, it is important to note that all the advantages of clean water, sufficient oxygen levels, etc. in a recirculation system have a positive effect on survival rate, fish health, etc., which in the end gives a high quality product.

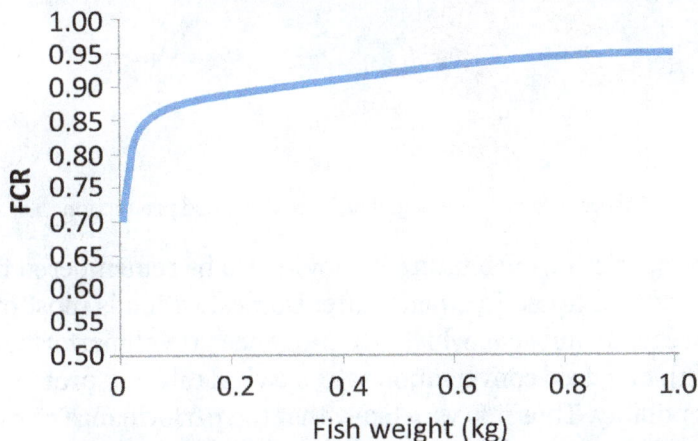

Figure: Feed conversion rate (FCR) of rainbow trout in a recirculation system,
related to fish weight at 15-18 degrees Celsius

Compared to other farmed animals there is a large variety of fish, and many different fish species are farmed. In comparison, the market for pigs, cattle or chicken is not diversified in the same way as fish. The consumer does not ask for different species of pigs, cattle or chicken, they just ask for different cuts or sizes of cuts. But when it comes to fish, the choice of species is wide, and the consumer is used to choosing from a range of different fish, a situation which makes many different fish species interesting in the eyes of any fish farmer. Over the past decade some hundred aquatic species have been introduced to aquaculture and the rate of domestication of aquatic species is around hundred times faster than that of the domestication of plants and animals on land.

Looking at the world production volume of farmed fish, the picture is not in favour of a multi species output. From figure below it can be seen that carp, of which we are only talking of some 5 different sub-species, is by far the most dominant. Salmon and trout are next in line, and this is only two species. The rest amounts to some ten species. One therefore has to realise that although there are plenty of species to be cultured, only a few of these go on to become real successes on a world-wide scale. However, this does not mean that all the new fish species introduced to aquaculture are failures. One just has to realise that the world production volume of new species is limited, and that the success and failures of growing these species depend very much on market conditions. Producing a small volume of a prestigious fish species may well be profitable as it fetches a high price. However, because the market for prestigious species is limited, the price may soon go down if production and thereby availability of the product rises. It can be very profitable to be the first and only one on the market with a new species in aquaculture. On the other hand, it is also a risky business with a high degree of uncertainty in both production and in market development.

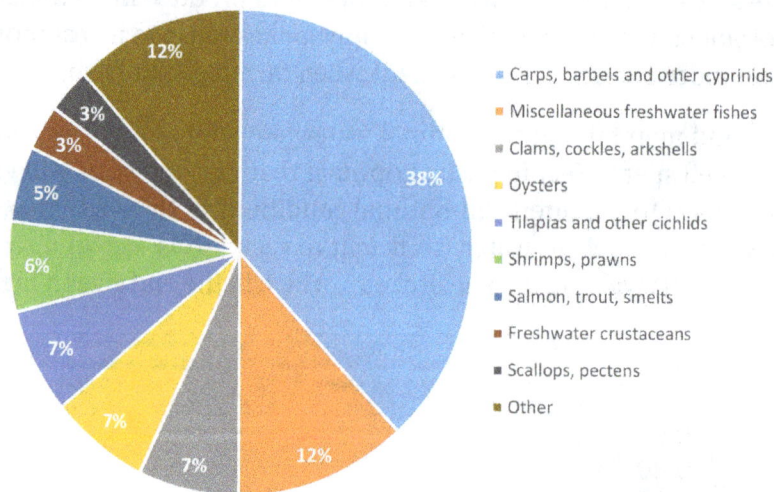

Figure: Distribution of global farmed seafood production

When introducing new species in aquaculture it should also be remembered that it is wild species, which are being captured and tested in aquaculture. Domestication is most often a long and troublesome task. There are many impacts, which will influence growth performance, such as high genetic variation in growth rate, feed conversation rate, survival rate and problems with early maturation and disease susceptibility. Thus it is very likely that the performance of fish from the wild does not correspond to the expectations of the aquaculturist. Also, viruses in wild stocks can be brought in, of which some only appear after several years breeding, resulting in a demoralising experience.

To give general recommendations on which species to culture in recirculation systems is not an easy task. Many factors influence the success of a fish farming business. For example, local building costs, cost and stability of electricity supply, availability of skilled personnel, etc. Two important questions though should be asked before anything else is discussed: does the fish species being considered have the ability to perform well in a recirculation facility; and secondly is there a market for this species that will fetch a price high enough and at volumes large enough to make the project profitable.

The first question can be answered in a relatively simple manner: seen from a biological point of view, any type of fish reared successfully in traditional aquaculture can just as easily be reared in recirculation. As mentioned, the environment inside the recirculated fish farm can be adjusted to match the exact needs of the species reared. The recirculation technology in itself is not an obstacle to any new species introduced. The fish will grow just as well, and often even better, in a recirculation unit. Whether it will perform well from an economic point of view is more uncertain as this depends on the market conditions, the investment and the production costs and the ability of the species to grow rapidly. Rearing fish with generally low growth rates, such as extreme cold water species, makes it difficult to produce a yearly output that justifies the investment made in the facility.

Whether market conditions are favourable for a given species reared in a recirculation system depends highly on competition from other producers. And this is not restricted to local producers; fish trading is a global business and competition is global too. Trout farmed in Poland may well have to compete with catfish from Vietnam or salmon from farms in Norway as fish are easily distributed around the world at low cost.

It has always been recommended to use recirculation systems to produce expensive fish, because a high selling price leaves room for higher production costs. A good example is the eel farming business where a high selling price allows relatively high production costs. On the other hand, there is a strong tendency to use recirculation systems also for lower priced fish species such as trout or salmon.

The Danish recirculation trout farm concept is a good example of recirculation systems entering a relatively low price segment such as portion sized trout. However, it is necessary for such production systems to be huge, operating in volumes from 1 000 tonnes and upwards, in order to be competitive. In the future, perhaps in some areas growing large salmon will move from sea cage farming to land-based recirculation facilities for environmental reasons. Even an extremely low priced fish product such as tilapia will probably become profitable to grow in some kind of recirculation system as the fight for water and space intensifies.

The suitability of rearing specific fish species in recirculation depends on many different factors, such as the profitability, environmental concerns, biological suitability. In the tables below fish species have been grouped into different categories depending on the commercial feasibility of growing them in a recirculation system.

It should be mentioned that for small fish the use of recirculation is always recommended, because small fish grow faster and are therefore particularly suited to a controlled environment until they have reached the size for on- growing.

Good biological performance and acceptable market conditions make the following fish interesting for production to market size in recirculation aquaculture:

Species	Current status	Market
Arctic char (Salvelinus alpinus) 14°C	Arctic char or cross breeds with brook trout has a long track record of growing well in cold water aquaculture.	Sold in specific markets at fair to good prices.
Atlantic salmon, smolt (Salmo salar) 14°C	Small salmon are called smolt. They are grown in freshwater before transfer to saltwater for grow-out. Smolts are raised in recirculation systems with great success.	The market for salmon smolt is usually very good. Demand is constantly increasing.
Eel (Anguilla anguilla) 24°C	Proven successful species in recirculation. Cannot reproduce in captivity. Wild catch of fry (elvers) is necessary. Considered threatened species.	Limited market with varying price levels. Some buyers will refuse to buy because of threatened species status.
Grouper (Epinephelus spp.) 28°C	Saltwater fish grown primarily in Asia. Many different grouper species. Requires knowledge in spawning and larval rearing. Grow-out relatively straight forward.	Sold primarily in local markets at good prices in areas where production by many small producers is taking place.
Rainbow trout (Oncorhynchus mykiss) 16°C	Easy to culture. Recirculation in freshwater widely used from fry rearing up to portion size fish. Larger trout can also be grown in recirculation whether fresh or saltwater.	Relatively tough competition in most markets. Products need to be diversified.
Seabass/ Seabream (Dicentrarchus labrax / Sparus aurata) 24°C	Saltwater aquaculture fish in a highly developed cage farming industry. Larval phases require good skills. Proven to grow well in recirculation.	Generally tough market conditions, but can fetch good prices for fresh fish in some local areas.
Sturgeon (Acipenser spp.) 22°C	Group of freshwater fish of many species relatively easy to culture. Skills required in different biological stages. Farming in recirculation systems is increasing.	Fair market conditions for meat. The caviar business seems to expand in high-end markets.

Species	Current status	Market
Turbot (Scophthalmus maximus) 17 °C	Good skills required in broodstock and hatchery management. Grows very well in recirculation.	Generally tough international market conditions. Local market prices can be higher.
Whiteleg shrimp (Penaeus vannamei) 30 °C	Most common shrimp species in aquaculture. Grow-out in recirculation systems has been proven successful. The production method is developing.	Shrimp prices are generally good and high in comparison to fish prices.
Yellowtail amberjack (Seriola lalandi) 22 °C	Yellowtail amberjack, or kingfish, is a saltwater species proven to perform well in cages and in land based systems.	Market prices good. Sold in specific markets.

Low market prices make the following fi challenging to produce with a profit in recirculate aqua-culture, and good marketing and sales efforts are important:

Species	Current status	Market
African catfish *(Clarias gariepinus)* 28 °C	Freshwater fish very easy to culture. A robust and fast growing fish that performs well in recirculation. Production must be very cost efficient.	Moderate to low prices. Most fish are sold live in local markets. Strong marketing effort required.
Barramundi *(Lates calcarifer)* 28 °C	Also called Asian seabass. Lives in both fresh and saltwater. Requires knowledge in larval rearing. Relatively straight forward in grow-out.	Sold primarily in local markets at fair prices. International market expected to grow as global marketing increases.
Carps *(Cyprinus carpio)* 26 °C	All carp species will grow very well in recirculation aquaculture systems. Keeping down production costs at a minimum is the main issue.	Carps are regarded as a low price species in most markets, but can fetch higher prices in some markets.
Pangasius *(Pangasius bocourti)* 28 °C	This catfish is grown in big earth ponds primarily in Vietnam. Impressive ability to survive and grow at sub-optimal conditions.	Low end product in the global fish market leaves no room for production costs.
Perch *(Perca fluviatilis)* 17 °C	Freshwater fish proven to grow well in recirculation although not widely used.	Limited market with fluctuating prices.
Tilapia *(Oreochromis niloticus)* 28 °C	One of the most predominant aquaculture fish, which is robust and fast growing. Production cost must be kept to a minimum to be competitive.	Sold in the world market at low to moderate prices. Can fetch higher prices locally.
Whitefish *(Coregonus lavaretus)* 15 °C	Coregonus is a group of freshwater fish that can be grown in aquaculture and in recirculation systems.	Prices relatively low as there is strong competition from wild caught species.

Very challenging to grow these fish at a commercial viable scale in recirculation aquaculture or in aquaculture in general, because it is either difficult to manage biologically or/and because of tough market conditions:

Species	Current status	Market
Atlantic cod (Gadus morhua) 12 °C	Fry rearing proven to be successful in recirculation. Grow-out of larger cod needs further development and is as such not suited for recirculation.	Prices are fluctuating as market is heavily affected by wild stock catches.
Atlantic salmon, Large (Salmo salar) 14 °C	Larger salmon are grown in sea cages to reach market size of 4-5 kilos. Grow-out in land based systems using recirculation is under development.	Global market dominated by Norwegian marketing. Trend towards certified products.
Bluefin tuna (Thunnus thynnus) 24 °C	Fattening of wild caught fish is the only profitable farming technology. Controlling full cycle at a commercial level in aquaculture is still under development.	Can fetch very high prices in a turbulent worldwide market for tuna.
Cobia (Rachycentron canadum) 28 °C	Fairly new saltwater aquaculture fish of good meat quality. Grow-out in cage culture. Output seems to be growing, although there are many obstacles in breeding.	Market is not well developed and the fish is unknown in most markets.
Lemon sole (Microstomus kitt) 17 °C	Not yet fully developed new species in aquaculture due to different obstacles in biology, such as feeding, etc.	High-end product fetching stable and high prices.
Pike-perch (Sander lucioperca) 20 °C	Freshwater fish difficult to farm. Larval stage troublesome, grow-out seems a little easier. Only a few successful recirculation systems for pike-perch.	Good and fair prices. Demand expected to grow as wild stocks fall and consumption increases.

Fish Slaughter

Slaughter is the term used when animals are killed for food. The methods used to slaughter fish depend on many factors, including fish species and size, fish numbers, type of aquaculture production system (e.g. ponds, cages, tanks), animal welfare concerns, regulatory compliance, economics, market preferences and effects on product quality. Some methods may not be approved in some countries, such as the use of certain anesthetics and other methods to kill fish.

Several organizations have prepared guidelines for humane treatment prior to and during slaughter, and slaughter methods may soon be specified in the purchase requirements of major seafood buyers. All methods of slaughter have issues where animal welfare concerns should be addressed. Approaches such as bleeding without stunning have been cited as less acceptable. Meanwhile, electrical stunning followed by an acceptable killing method, a combination that minimizes physical and biological effects, may have less animal welfare issues.

Selecting a Slaughter Method

Prior to slaughter, feed is withdrawn from cultured fish to allow their digestive systems to empty their contents. This reduces the chance of fish tissues being contaminated by gut contents during processing and helps to maintain the quality and hygiene of the final products.

Fish should not be crowded. Crowding exposes fish to a rapidly increasing density, and as a result, oxygen availability and water quality can decrease rapidly. Overly crowded fish may show vigorous agitation, gasping and splashing. There is also an increased risk of damage from abrasion through contact with other fish, nets or holding tanks. The adverse effects of crowding can be lessened by reducing densities slowly and providing additional oxygen. Fish should be crowded for as short a period of time as possible.

Harvesting fish for slaughter usually requires some handling and concentration of fish, which creates stress for the fish. Harvesting can cause elevated levels of cortisol, which is the primary stress hormone in fish, as well as lactic acid and glucose. Stress can also cause reduced glycogen levels, decreased muscle pH and rapid onset of rigor mortis. Thus, undesirable pre-harvest physiological changes that stress the fish can result in lower product quality and reduced processing yields. This can have a significant effect on profitability.

Slaughter Methods

A variety of slaughter methods are currently used for fish, depending on the species, desired product quality and market demand. Some fish are killed and processed individually, while others are killed and processed collectively. Slaughter is generally a twostage process. First, the fish are stunned so they cannot sense pain. Then the fish are killed by one of a variety of methods. It is important that the stun-to-kill time is brief so that fish do not regain consciousness before they are killed.

When choosing methods for slaughter, it is important to evaluate how each method will impact the fish, personnel, and resulting product. For each slaughter situation, be sure the method:

- Induces rapid loss of consciousness and death in fish.
- Creates a minimal amount of pain and distress.
- Produces irreversible loss of consciousness.
- Works with intended fish use and purpose.
- Provides reliable slaughter methodology.
- Works with species, size and age of fish.
- Is appropriate for the subsequent evaluation or use of fish tissues.
- Causes minimal environmental impact of fish carcasses be consumed.
- Is safe for personnel who use methodology.
- Minimizes emotional effect on observers or operators.

- Causes no damage to equipment.
- Meets regulatory requirements.

Fish Welfare and Common Slaughter Methods

Acceptable slaughter methods must render the animals insensible immediately and should be performed without causing undue pain or suffering. Rapid slaughter methods are preferred over slower methods, since they lead to improved fish welfare and fish quality.

Fish Slaughter Methods

Carbon Dioxide

Carbon dioxide saturated water causes narcosis and loss of consciousness in the fish. Only carbon dioxide from a source that allows for careful regulation of the concentration, such as from a gas cylinder, is acceptable.

The gas is dissolved in the water prior to fish being introduced to the water. Fish placed in this water often react violently while their blood absorbs the gas rapidly. The fish may get bruised from hitting other fish or the sides of the holding container. The time required for fish to be killed can vary from a few minutes to 10+ minutes. Thus, carbon dioxide is often used as the first step in a two-step process. First, fish are anesthetized by the carbon dioxide. Then, fish are killed rapidly by a secondary method. Some countries do not allow the use of carbon dioxide due to welfare concerns.

Captive Bolt

With the captive bolt method, fish are oriented into a machine individually that uses a pneumatic non-penetrating captive bolt to stun the fish. This technique is accurate and delivers the required concussion velocity to stun the fish quickly. Fish are then killed manually or mechanically by exsanguination using an acceptable bleeding technique, such as cutting or removing the gills.

Gunshot

This ballistic technique is used with large wild-caught fish species such as halibut and tuna. The bullet penetrates the brain, where it causes extensive damage and immediate death of the fish. Inaccuracy and human safety are major concerns of this technique.

Spiking

In spiking, pithing, coring or "ikejime", a spike or probe is inserted through the skull of the fish directly into the brain. The fish are stunned and killed at the same time. This procedure can be applied more accurately in large fish due to the larger size of the brain. In smaller fish, the brain may be difficult to locate and destroy. If the brain is not adequately destroyed, the fish can undergo distress, and undesirable tissue quality changes may result.

Other Manual Techniques

These methods include manual blunt force trauma, decapitation and cervical transection. With blunt force trauma, fish are removed from the water individually and given a sharp blow to the head. If the blow is accurate and strong enough, the animal is killed instantly. However, if the blow is weaker, the animal is only stunned and a second step of killing is required. Both decapitation and transection of the cervical spine also require a secondary kill method such as pithing.

Electrical Stunning

Stunning by use of electricity is known as electronarcosis. Killing by electricity is known as electrocution. Electrocution completely destroys brain function and renders the animal unconscious while stopping the breathing reflex, so the fish is stunned and killed simultaneously. Electric stunning is reversible, as normal brain function is disrupted for only short period of time. Hence, electronarcosis must be followed immediately by a secondary kill method. Electric stunning immobilizes the fish and prevents distress and struggling prior to slaughter, which are detrimental to tissue quality. Recent advances in electrical equipment design have made substantial improvements in preventing or minimizing undesirable physical and biological effects in electric stunned fish. For electrical stunning to be effective, proper current and stun duration must be maintained. Also, water factors such as conductivity and temperature must be properly managed.

Bleeding

Killing a fish by bleeding can be accomplished by any of three major processes: cutting the gills, removing the gills or severing the caudal vessels of the tail. Fish should be stunned prior to bleeding. Fish die of anoxia and proper stunning should prevent any distress or struggling of the fish. Bleeding prevents fish muscles from turning an undesirable red color or acquiring a blood odor, which can prevent fish from being sold for sushi or surimi.

Rapid Chilling

Placing fish in iced water can be used to immobilize fish so they can be more easily handled for a secondary killing method. Fish can also be placed in an ice bath for a longer time until death occurs, but this technique is only appropriate for warm water fish species and not recommended for temperate, cool or cold water species, or for medium to large-bodied fish where the time to death may be significantly prolonged. Even though fish usually struggle prior to loss of consciousness, live chilling is considered by many in the aquaculture industry to offer benefits to carcass quality, because reducing muscle temperature close to zero degrees Celsius helps delay enzymatic and microbial spoilage processes. The process also increases the time for onset of rigor mortis and the resolution of rigor. Another advantage of this slaughter method is that the water can be drained and the fish placed in an iced container with the fish's temperature already lowered.

Asphyxiation

Asphyxiation by removal from water, or "dewatering", has been commonly used by the aquaculture industry for many years. Basically, fish are removed from the water and allowed to die via asphyxiation. The fish struggles as oxygen is depleted and respiration ceases. The rate at which oxygen depletion occurs a depends upon ambient temperature and the rate of fish activity (e.g. reducing the temperature of fish typically prolongs the time to anoxia and, therefore, the time to insensibility, thereby lengthening the period of distress). This method is less acceptable compared to the afore-mentioned methods and is not considered acceptable by American Veterinary Medical Association.

Anesthesia

With the exception of carbon dioxide, the use of anesthetics is not an acceptable method for slaughter sas there are regulatory and human consumption concerns over potential anesthetic residues in the fish tissues.

References

- Torrissen, Ole; et al. (2011). "Atlantic Salmon (Salmo Salar): The 'Super-Chicken' Of The Sea?". Reviews in Fisheries Science. 19 (3): 257–278. doi:10.1080/10641262.2011.597890

- Development-of-intensive-fish-farming, farmed-fish-welfare: fishcount.org.uk, Retrieved 13 May 2018

- "Fish Farms Drive Wild Salmon Populations Toward Extinction". ScienceDaily. 16 December 2007. Retrieved 2018-01-06

- Integrated-recycling-systems: fishnetafrica.org, Retrieved 30 April 2018

- Avnimelech, Y; Kochva, M; et al. (1994). "Development of controlled intensive aquaculture systems with a limited water exchange and adjusted carbon to nitrogen ratio". Israeli Journal of Aquaculture Bamidgeh. 46 (3): 119–131.

- Fisheries-types-of-fisheries-and-it-economical-importance-1361: biologydiscussion.com, Retrieved 24 June 2018

- Huntingford, F. A; Adams, C; Braithwaite, V. A; Kadri, S; Pottinger, T. G; Sandoe, P; Turnbull, J. F (2006). "Current issues in fish welfare" (PDF). Journal of Fish Biology. 68(2): 332–372. doi:10.1111/j.0022-1112.2006.001046.x

- Fish-farming-105599: wellnessmama.com, Retrieved 14 March 2018

- Barrionuevo, Alexei (July 26, 2009). "Chile's Antibiotics Use on Salmon Farms Dwarfs That of a Top Rival's". The New York Times. Retrieved 2009-08-28

- Myths-indoor-fish-farming: harvestreturns.com, Retrieved 13 May 2018

- Ford, JS; Myers, RA (2008). "A Global Assessment of Salmon Aquaculture Impacts on Wild Salmonids". PLOS Biol. 6 (2): e33. doi:10.1371/journal.pbio.0060033. PMC 2235905. PMID 18271629

Chapter 4

Mariculture

Mariculture is an area of aquaculture, which involves the farming of marine organisms in the open ocean, in enclosed section of the ocean, tanks, raceways or ponds that are filled with seawater. An understanding of mariculture requires a study of the cultivation of algae, shellfish, etc. which have been extensively discussed in this chapter.

Mariculture is an activity involving food production for human consumption. It is an activity in which aquatic organisms both plants and animals are cultured in a confined environment in the aquatic medium which may be completely marine or marine mixed to various degrees with freshwater in the brackishwater areas.

Need for Mariculture

It is accepted that with world population increasing, the food available per head is diminishing and it is therefore necessary not only to intensify production from existing areas but also to find additional areas for food population. If the experience in the eastern countries is any guide it will be seen that the coastal swamps can be profitably diverted to food production and in this process gain other side benefits such as creating employment in the rural areas and at least to some extent diminishing the treck to urban areas with its associated problems. It has also been shown that production from a unit area of a controlled environment is much more than from open natural waters.

Structures for Mariculture Operations

Four main types of structures are in popular use at present:

1. Floating cages: In the open sea or large sheltered bays floating cages are used for culture purposes. Other structures are not practicable because of constant wave action. But for culture in cages because of high cost of cages it is necessary to maintain a large number of individuals inside the cage to make the venture profitable. This in turn necessitates supplementary feeding since the quantity of natural food available in the volume of water inside the cage cannot sustain the large number normally cultured in these cages. For this type of culture practice to get popular it will therefore be necessary to first develop inexpensive cages and also fish feed at reasonable prices and understandably as a result, this method has been slow to develop in developing countries.

2. Net enclosures: Net enclosures barricading off large areas in sheltered bays are being tried on a commercial scale in some countries. But this requires considerable capital outlay and frequent replacement as a result of corrosion. In addition, supplementary feeding is essen-

tial which requires higher running costs - but it does pay its way that is why some private farms are existing in developed countries.

3. Earth ponds: For inland brackishwater areas where the tidal range is adequate, experience in developing countries in the east has shown that for mariculture, earth ponds constructed to impound spring tide water is the most suitable structure. This can be done without much difficulty in areas where the land is more or less flat or has slight slope and the nature of the soil is satisfactory. The construction of earth ponds and use of tidal water encourages the use of manual labour for construction and harneses gravitational force in the form of tides for the water supply. It would not only be independent of any foreign imported energy source but also reduce the imported material component to negligible quantities.

4. Constant water circulation systems: Constant water circulation units are popular in some developed countries. These are large cement structures. In addition this requires continuous pumping of water in large quantities and also supplementary feed. Thus this system requires not only a heavy capital outlay but also high recurrent costs. This proves economical where a high price can be obtained for the end products and adequate raw material for feed is available.

Requirements for Coastal Mariculture Development in Earth Ponds

Water Supply

The most common source of water supply for coastal ponds is the tide. For this the tidal amplitude should be adequate for frequent changes of water in the ponds. For good results it is best to have 75 cm to 1 m of water above the ground at high tide and at low tide the water level should go down at least 75cm below the existing ground level in order to facilitate complete draining of the ponds. While these are the desirable levels nevertheless successful operations are carried out in areas with much lower tidal amplitude.

It is essential that the water supply should be polution free and preferably of pH ranging from 6.5 to 8. The depth of water required inside the pond would depend on:

 i.   The species to be cultured and

 ii. As to whether it is to be.

 a.   An extensive system i.e. depending entirely on the natural primary production within the pond for food or

 b.   Semi intensive using fertilizers or

 c.   Intensive system where the organisms cultured would be given supplementary feed.

Soil Conditions

For earth ponds availability of soil of good water retaining capacity is important in order to prevent loss of water by seepage specially when it is proposed to use fertilizers to increase production. It is also essential that the pH is within the desired limits as for the water given above. In addition acid sulphate soils should be avoided.

Seed for Stocking

Availability of adequate quantities of seed of the species to be cultured is an important factor. If adequate quantities are naturally available it would reduce the cost of production considerably. Hatchery production of seed is being successfully carried out for some species in other countries. This may be considered only when the natural supply is inadequate and demand for hatchery produced seed is adequate.

Culture Methods in Earth Ponds

There are three main methods, which also represents gradual development in complexity both in management and financial outlay and of course coupled with higher production.

Extensive

This is the simplest form of culture where no fertilizer or supplementary feed is added and the production from the pond depends entirely on the primary production which in turn depends on sunlight and nutrients available in the water and soil. The stocking density will therefore have to be low and the recurrent costs are the lowest. For this method, in the case of mullets, milkfish and other herbivores or penaeids it is essential that the water level be maintained between 35 and 40cm in the general platform area of the pond to encourage algal growth on the pond bottom for the organisms to feed upon. If the water is deeper the light intensity that reaches the bottom is not sufficient to promote algal growth. In the tropics if the water level is reduced below 35 or 30cm then the temperature is likely to increase and may reach lethal limits. In rural areas of developing countries this method is popular mainly due to the low financial outlay required for the operation. This method is also useful in the initial stage of mariculture development in developing countries where it is still new and trained manpower is inadequate. Since an initial training phase is necessary and during this phase improper operations can lead to loss of an entire harvest. Extensive systems because of the low financial outlay can better withstand these shocks than the other two systems. In addition adequate experience with the basic system is very helpful in understanding the more complicated systems and helps in mastering the complicated system faster.

Semi-intensive

The next step towards increased production is to stock at a slightly higher density and then to add fertilizers (organic or inorganic) and in turn increase the harvest from the pond. The function of the fertilizer is to increase primary production within the pond which starts the chain reaction and ends in increased production from a unit area of the pond. This method not only requires a slightly higher financial operational cost, but also involves additional labour for the collection, storage and application of fertilizer. In developing countries, experienced farmers often fertilize their ponds because of the increased profit margin.

Intensive

In this method supplementary feed is added to the pond with or without the use of fertilizers. The stocking rates would be much higher and also the depth of water in the pond greater than in the former cases because the food for the cultivated organisms is not dependent entirely on primary production within the pond. However, feed is expensive and as a result the operating costs are much higher. The higher rewards also carry with it greater risks. Any unforseen calamity can wipe out a big financial outlay. For this reason and considering the heavy financial outlay required, at the initial stages where coastal aquaculture is still in its infancy and experience is inadequate, commencing with an extensive system would involve much lower risks and establish greater confidence.

Algae

Algae farms are places where algae are grown for commercial use. People engaged in algae farming are said to be involved in algaculture. Algaculture can involve growing many different species of algae. Most types of algae that are commercially grown are microalgae. These are sometimes referred to as phytoplankto, microphytes, and planktonic algae. Some of the larger algae species, also known as macroalgae, include seaweed and also have commercial uses.

However, their larger size and the more precise growing conditions for macroalgae make them more challenging to cultivate in a controlled environment. Some of the uses of commercial algae grown on algae farms includes food coloring, fertilizer, bioplastics, chemical feed, medicines, pollution control, and fuel. Some varieties are also grown as food for humans or as nutritional supplements.

Most algae farms grow only one type of algae. This is known as monoculture. Algae farmers choose the variety they are going to grow and take great care to keep their supply pure, as it is easy for other species to get into an algae culture and then come to dominate it. Pure cultures of one type of algae are the most valuable for commercial purposes and for research.

When mixed algae species are grown together, it is usually as food for other sea creatures, such as larval mollusks. This is a relatively low maintenance way to feed commercially farmed seafood.

All that is required is a filter to remove algae that is too large for seafood to eat, and a day or so in a greenhouse pool or outside so the algae gets the necessary nutrients to make it healthy for the seafood it will feed. When cultivating algae on algae farms, there are some basic requirements for producing a good stock, regardless of the species. Light, water, and minerals are all important ingredients in producing healthy algae. So is carbon dioxide. This combination produces the energy algae needs to grow.

Algae can sometimes be grown without sunlight if sugar is used to directly feed it. In these cases, carbon dioxide is also not required to grow the algae. Most algae farmers prefer the light method.

The temperature of the water in which the algae is grown is also important. Each species of algae has its own particular temperature range in which it does best. However, the optimum temperature range for algae of any kind is between 25 and 35 degrees C.

When it comes to harvesting algae, several different methods can be used. The most common are flocculation, centrifugation, and microscreening. Flocculation is an expensive method of harvesting algae that only large algae farms can usually afford.

It uses the powdered shells of crustaceans to interrupt the carbon dioxide supply of the algae, which causes algae to float to the surface of the water where it can be skimmed.

Centrifugation spins the water containing the algae in a centrifuge to separate the algae from the water and is a medium-cost method of harvesting due to the cost of the centrifuge. Finally, microscreening simply uses a fine mesh screen to sift the algae directly out of the water.

From harvesting, the algae is then dried and sent to the various companies around the world that have ordered it for the many purposes it can serve.

Algae farming promote clean technology. It is a ready source of food for both animals and humans. It can be used as nutritional supplements almost as-is. It has the ability to clean up pollution from the water and the air.

Most of all, it provides a clean source of fuel for transportation and heating that produces no greenhouse gasses or other pollution when it's used. It's non-toxic and safe to grow and harvest. It is also an extremely renewable resource.

Algae have been around for millions of years and will probably be here as long as there is a habitable earth. We can't run out of it. This makes algae farms part of the green technology of the future.

Shellfish

Shellfish aquaculture is the farming (i.e., cultivation and harvest) of aquatic invertebrates, such as oysters, clams and mussels. Cultivation implies involvement in the rearing process to enhance production, such as regular stocking and protection from predators. Shellfish farming has been a part of British Columbia's history for over 100 years. The systems used to farm shellfish have evolved from purely beach to technology-based systems that are designed for specific species and farming sites.

Shellfish begin their lives as larvae that mature into seed and/or juvenile animals. The farm cycle begins with the collection of larvae, which may be gathered in the wild or produced from hatchery broodstock (depending on the species and location).

- Clam larvae are kept in hatchery tanks where they transform into seed.

- Mussel larvae transform to juvenile animals.

- Scallop larvae settle and become juvenile animals.

- Oyster larvae are kept suspended in tanks by circulating water until they transform into seed.

Farmers acquire clam and oyster seed at various stages of its development – depending on the requirements of the tenure and farming operations. The seed is put into a nursery environment where it is nurtured into juvenile animals. Generally speaking, the juvenile animals then graduate to the growout phase of their development.

- Clams are spread on subtidal tenures where they burrow and mature to marketable size over a period of two to four years.

- Mussels are relocated to deepwater tenures where they are suspended in mesh socks to mature to marketable size over a period of 18 to 36 months.

- Scallops are transferred to deepwater tenures where they are suspended in a mesh bag or tray (suspension culture) or are seeded on the ocean floor (bottom culture). Maturation to marketable size takes six to 36 months in suspension culture and an additional 24 to 36 months for bottom culture.

- Oysters are frequently moved to a floating upwelling system (called a flupsy). Ocean water is circulated through the flupsy and juvenile animals, kept in trays, are able to grow to a larger size. When they are large enough, the young oysters are moved to be reared in a growout system. The most common growout techniques are raft, longline and intertidal.

Raft System

Shellfish farmers use rafts at deepwater sites to suspend culture systems used for different stages in the rearing of oysters, clams and mussels. For example, many shellfish farmers suspend tray systems that are used as nurseries for juvenile oysters and clams as well as for oyster growout. This method of "off bottom" farming is considered one of the most productive in the world.

Rafts must be built to withstand the severest weather, hold the weight of hundreds of dozens of mature oysters, and serve as safe work platforms for workers handling product. Important considerations in the design of a raft system include flotation, flexibility, stability, functionality, durability and capacity.

The raft system must be securely anchored to prevent movement and/or drift. Rafts are usually roped together and securely tied at three points on each raft and then anchored at each end. Anchor ropes will sway in the currents and slacken at low tides.

Longline System

Longlines are used worldwide to grow everything from scallops to seaweed. These systems are preferred in high exposure areas. They are productive and flexible enough to handle a variety of shellfish species as well as a range of culture systems. In a longline system, the farmer anchors a length of line at both ends, attaches a flotation and hangs culture systems on the line.

Nursery rearing as well as grow-out can be accomplished on longlines. Trays, tube modules and bags or cages can be hung in deep water for nursery rearing of clams, oysters or scallops. Seeded lines or socks (with adequate predator protection) are commonly suspended from longlines. Scallops are frequently grown-out on sunken longlines, in suspended lantern nets, or ear-hung directly on a down-line.

Layout of a longline system depends on site characteristics. The most significant feature, from a security and stability perspective, is availability of shore to anchor one end of the longline. In some locations, both ends can be fixed to the shore. Anchoring both ends in deep water may be done at sites where shore anchoring is not possible or desirable.

Intertidal System

Intertidal farming refers to systems in which shellfish are exposed to air during the low tide of each tidal cycle. These systems include both bottom (beach) and near-bottom (epibenthic) techniques.

Managing and maintaining productive intertidal growing areas is no different than land agriculture. The substrate will be cleared and prepared for planting. Both oyster and clam farming may require substrate improvement to reach acceptable levels of productivity. The area will be seeded and, in many cases, the seeded plots are protected from predators by overlaying them with mesh that is then secured into place. The plots will be tended regularly. Farmers will develop and maintain an inventory control system to know what was planted when and how it is performing.

- Clams perform best in a substrate composed of a mixture of mud, sand, pea gravel and some shell fragments. Improvement for silted beaches may mean gravelling while for other sites it involves debris and rock removal.

- In oyster farming, intertidal grow-out systems include beach distribution of seeded shell cultch as well as oysters grown on stakes, racks and intertidal longlines. Near-bottom methods have been adopted on sites where bottom conditions are not suitable for growing oysters; e.g., soft mud, silt. Nursery rearing of oysters on shell cultch or tubes may also be done intertidally.

Shellfish culture is a major sector of aquaculture production worldwide, and zoonoses and drug residues associated with shellfish farm practice are of concern to public health. Although many diseases can affect shellfish, they do not appear to be transmittable to humans. Rather, the main hazards are associated with the methods used to farm the different species. The risk to human health from shellfish most commonly relates to contamination by biotoxins produced by marine algae. Another well-recognised problem associated with shellfish culture is the contamination of shellfish with domestic sewage that contains human pathogenic bacteria and viruses, which causes diseases such as typhoid fever and hepatitis. In shrimp farming, the main potential food safety hazards are zoonoses, chemical contamination and veterinary drug residues. Untreated effluent from shrimp farms is a major concern to the environmental sector as it is known to promote plankton blooms if directly discharged into natural water sources.

Shellfish are filter feeders and as such they filter plankton (microscopic green plants) from the seawater converting it to wholesome animal protein. Their shape, the fullness and texture of the meat, and their flavor are determined by where they grow, how they are farmed and the plankton they eat.

Ecological Services of Shellfish Farming

1. Shellfish clean the water by filter feeding. We estimate that the oysters and mussels growing at Salt Water Farms clear 25 million gallons of bay water each day. This improves water quality by reducing turbidity, improving light penetration and reducing anoxia (low oxygen).

2. Shellfish convert nitrogen into edible animal protein. When harvested the nitrogen is permanently removed from the ecosystem. Excess nitrogen in coastal waters contributes to anoxia and kills fish.

3. Shellfish farms provide essential habitat for a wide variety of marine plants and animals.

4. Shellfish farms provide traditional marine related jobs in the local community.

5. Shellfish farms provide wholesome and tasty farmed goods to the restaurants that feature local foods.

Artificial Reefs

Artificial reefs are designed to enhance recreational fishing opportunities by providing additional habitat for fish and other aquatic organisms, increasing their numbers in the area.

The reefs come in many different sizes, shapes and models, for example some are tower-like steel structures whereas others are cube-shaped, constructed from reinforced concrete. Once installed on the ocean floor the reef structures begin to be populated by aquatic life, attracting more organisms and fish, and a food web begins to develop.

The main purpose of artificial reefs is to act as a dissuasion tool for fisheries (i.e. illegal trawling) and as a protection tool for marine resources, environment and other legitimate activities.

This application is frequently used to protect habitats of ecological interest or of importance for some life stages of some resources (e.g. Posidonia beds, maerl beds, coralligenous, biogenic reefs, reproduction and nursery areas, sensitive and essential fish habitats, etc.) from illegal trawling, dredging and bottom purse-seining that can damage both the habitat and its associated resources. The use of appropriately-designed artificial reefs may help control and reduce conflict between trawling and coastal small-scale fisheries using set gear.

Some protection artificial reefs can be used to protect other structures like cables, oil or waste water pipelines thereby preventing pollution damage.

Design and Material

Protection artificial reefs should be specifically designed to withstand the power of fishing vessels in an area and to either hook nets or tear them up. Therefore, the units must be heavy enough to steadily maintain their position on the seabed and avoid removal by fishing vessels. Several artificial reefs have failed because the modules were shifted or hauled up by the fishing vessels. Consequently, protection units should be dense and relatively low profile, with a low volume in relation to their weight. The weight should be related to the power of the fishing vessels to be stopped. Concrete blocks with deterrent arms are usually employed.

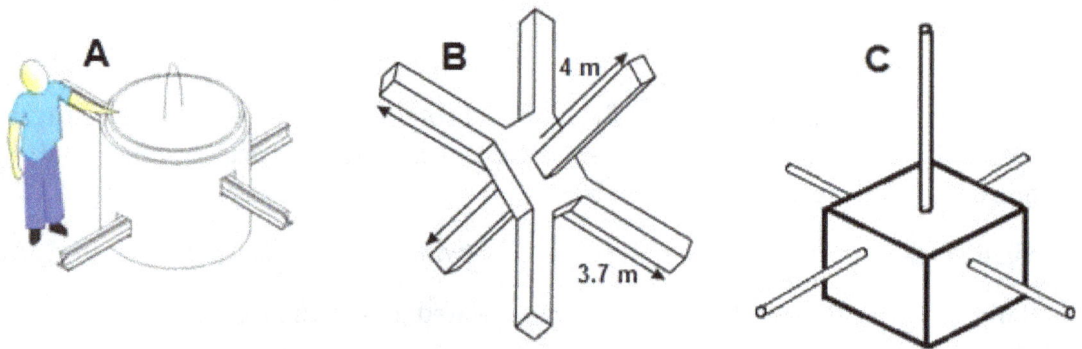

Figure: Examples of protection units

Figure shows the technical parameters to be considered in designing protection artificial reef units. Considering the extended area typically covered by protection artificial reefs, planning the location of the units on the seabed requires detailed knowledge of several features of the seafloor such as the distribution of natural habitats and the position of man-made structures (pipeline, cables, etc.) in order to protect them, avoid damages and prevent negative impacts.

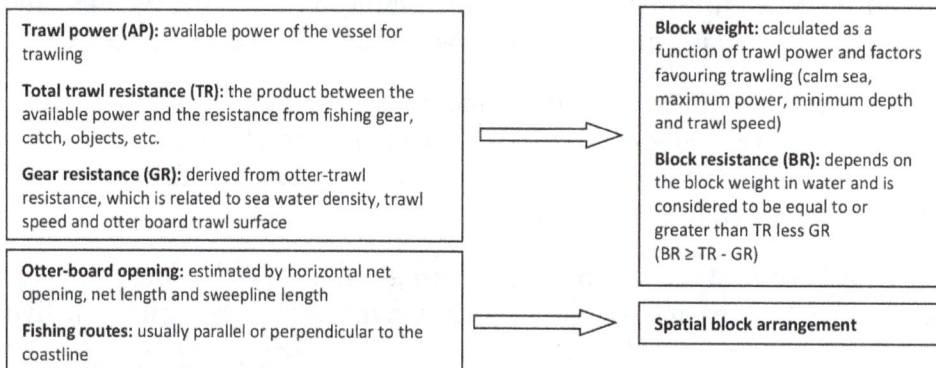

Trawl power (AP): available power of the vessel for trawling

Total trawl resistance (TR): the product between the available power and the resistance from fishing gear, catch, objects, etc.

Gear resistance (GR): derived from otter-trawl resistance, which is related to sea water density, trawl speed and otter board trawl surface

Otter-board opening: estimated by horizontal net opening, net length and sweepline length

Fishing routes: usually parallel or perpendicular to the coastline

Block weight: calculated as a function of trawl power and factors favouring trawling (calm sea, maximum power, minimum depth and trawl speed)

Block resistance (BR): depends on the block weight in water and is considered to be equal to or greater than TR less GR (BR ≥ TR - GR)

Spatial block arrangement

Figure: Variables to be considered when designing anti-trawling reef units

It is also essential to know the fishing routes in the area in order to strategically place modules along designated lines perpendicular to the routes. The distance between modules should be less than the otter-board/dredge openings, hence than the free space needed by the vessel to pass with the towed gear between one module and the other, taking into account the best relationship between the artificial reef's effectiveness and costs. Usually, these modules are placed alternate along two or three parallel lines.

When protection artificial reefs are deployed to create suitable grounds for selective small-scale fisheries and to protect the resources from other less selective fishing activities, the reef units should be placed following a spatial design which allows for the use of set gear within the reef area.

Several protection artificial reefs have failed in their protection function because the units were haphazardly dropped from the sea surface and, hence, became scattered on the seabed without following a specific design. Therefore, the use of GPS combined with controlled release of the modules by crane can assure their correct positioning and effectiveness.

Practical Applications

Several examples of this application exist in the Mediterranean Sea (e.g. Spain, Tunisia).

Spain

The development of artificial reefs in Spain is motivated by the need to protect coastal fishing resources, high-diversity biological communities and selective small-scale fisheries against the action of non-selective fishing methods like trawl and seines.

More than 130 artificial reefs have been constructed along the Spanish coasts since 1989, most of them for protection purposes, as a tool for Spanish fisheries policies. Along the Mediterranean coast, the depth of deployment ranges from 10 to 35 m – sometimes up to 50 m. The projects developed in Spain for protection artificial reefs have tried to optimize the design of units to improve their function and optimize both the number of units and their arrangement on the seafloor. The goal was to protect as much area as possible minimizing costs and habitat modifications. Cost reduction was also achieved using maritime conventional means to install the reefs without the intervention of divers.

The results indicate an increase of local fishing resources, a reduction of conflicts between fishers and, in some cases, a significant recovery of natural habitats.

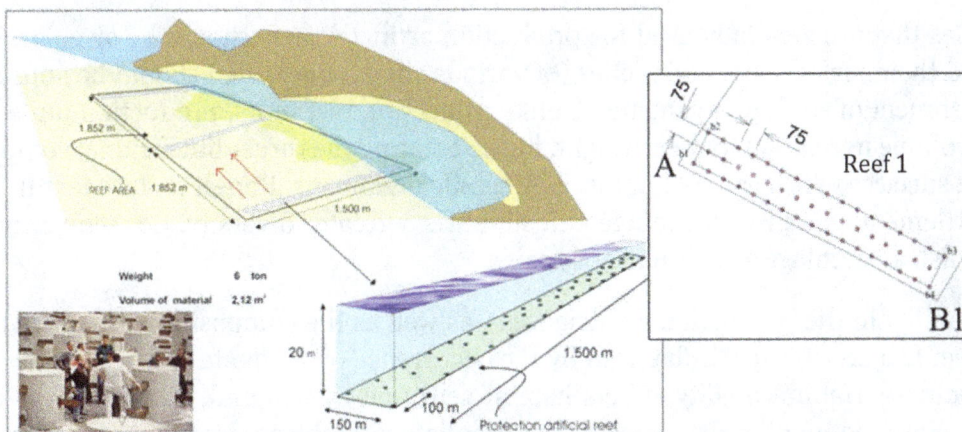

Figure: A protection artificial reef (Spain). The protection units are placed along three lines perpendicular to trawlers navigations routes (red arrows) to protect the Posidonia and Cymodocea beds (green area) inshore and leave space for artisanal fishing activities

Production Artificial Reefs

The overall objective of production artificial reefs is to increase the productivity of the aquatic environment and promote a sustainable utilization of resources.

When opportunely designed, artificial reefs may increase the biomass and hence the availability for human consumption of a variety of aquatic organisms (algae, molluscs, sea-urchins, fish) by enhancing their survival, growth and reproduction providing them with suitable habitats and additional food.

This type of artificial reef can also be used to manage the life stages of targeted species favouring aggregation of juveniles in certain areas and gathering the adults at suitable fishing grounds.

The specific applications of production artificial reefs include:

- Recovery of depleted stocks by increasing juveniles survival providing them with shelter and additional food;

- Enhancement of local fisheries by aggregating and establishing permanent populations of fish at suitable fishing grounds;

- Shifting the fishing effort from an overexploited resource to other resources; e.g. if the soft-bottom associated species in an area are overexploited, artificial reefs can serve to shift a part of the fishing effort to pelagic or reef-dwelling species;

- Compensation for a reduction of fishing effort: when there is a need to reduce fishing effort of trawling in an area, production artificial reefs can be used in negotiation to create new fishing grounds allowing fishers to shift towards more selective fishing activities;

- Development of extensive algae and molluscs aquaculture, providing suitable substrates for settlement.

Design and Materials

The modules that are generally used for production artificial reefs should be alveolar, of various shapes, and their surface area and niches (of various shapes and sizes) should be appropriate for the establishment of settling organisms. Unlike protection reef units, production units have usually more volume in relation to their weight, hence creating the three-dimensional complexity and developing surfaces which can be colonised by sessile organisms. Rough surface texture enhances benthic settlement as it provides refuge and supports a greater diversity. Consequently, this also affects the fish assemblage attracting fish grazing.

Food availability in the production artificial reef, as well as the composition, diversity and abundance of reef fish are strongly influenced by the occurrence of adequate refuges and by the shape of the structures. Habitat quality affects habitat selection by fish and, consequently, influences demography and population dynamics of the reef fish assemblage. Hence, to host a permanent community, an artificial reef must provide adequate habitats to juveniles and adults. On the basis of the fractal crevices theory in structurally complex natural or artificial environments, large crevices are much rarer than the smaller ones. Therefore, the artificial reefs can host a greater number of small and medium-sized organisms than large ones which tend to migrate outside. Therefore, the placement of large- holed reef units (especially in MPAs) could avoid depletion of broodstock by fishing and enhance the reproductive capacity of reef fish.

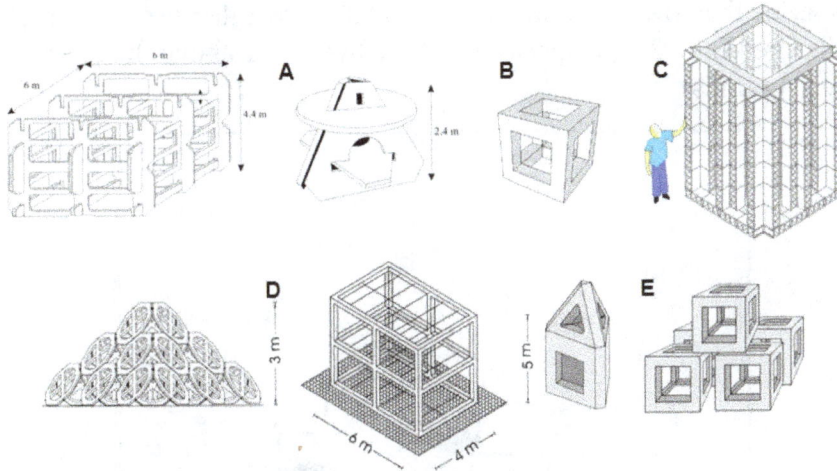

Figure: Examples of production artificial reef modules

Other factors that should be taken into account in planning artificial reef structures are:

- Regardless of their size and life stage, fish generally prefer cavities where there is light and with many openings that enable them escaping from predators;

- The size, number and orientation of cavities should match with the behavioural features of the target species, such as whether they are territorial or gregarious;

- The overall design of artificial reef structures should assure adequate water circulation;

- If demersal species are targeted, the structures design should consider providing vertical cover such as an "overhang" or shading from above as demersal fish will more readily utilize structures which provide protection from predators foraging from above.

With regard to the shape of the reef units/reef sets, it is well known that the affinity of several aquatic organisms towards the artificial substrates vary widely depending on the species and the life stage. Because of this, when constructing a reef for fisheries enhancement, it is important to deeply know the ecology of the different species so to identify those that are more appropriate as targets for artificial reef deployment and will have a higher probability of being manageable through manipulations involving artificial reefs.

Fish species have been classified according to their affinity to artificial reefs:

- Type A: benthic, reef-dweller organisms (fish, crustaceans, cephalopods) that prefer to live at strict contact with the substrates or inside holes (e.g. gobids, blennids, scorpenids, octopus and lobsters);

- Type B: nekto-benthic, reef-dweller fish that swim around the structures but are linked to them by the occurrence of shelter and/or prey availability (e.g. sparids, sciaenids, seabass and labrids);

- Type C: pelagic fish swimming in the middle and surface layers of the water column; they usually maintain a certain distance from the artificial structures but are likely to be linked to them by vision and sounds (e.g. mugilids, amberjacks and dolphin fish);

- Type D: species that are found on, in, or over the substrate next to the reef. These species have similar needs to C-type species but they live on or above the substrate surrounding the reef (e.g. bothids).

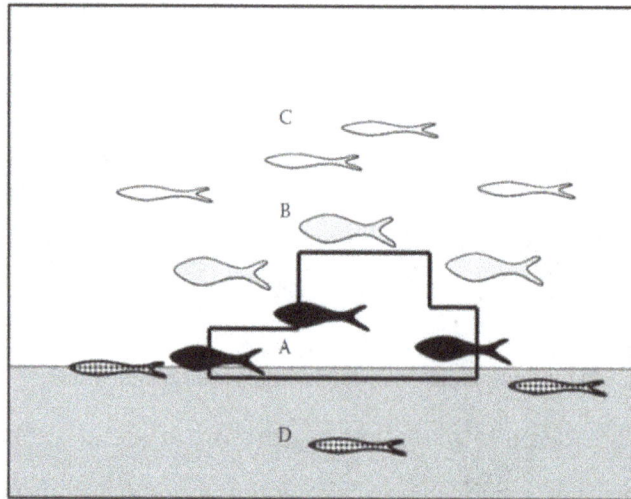

Figure: Classification of fish according to their position relative to the artificial reef

In figure the different fish categories are displayed along two axes (attraction and production) according to their level of affinity towards hard substrates. C and D-type species are characterized by high attraction and low production relationships with artificial substrates, hence they are clearly not suitable to be managed with an artificial reef in terms of increasing production as these species are chiefly attracted to the reef. A-type species, which have a strong production relationship (e.g. spiny lobster or octopus) might gain a significant advantage from artificial reef deployment, while B-type fish will get benefit from artificial reefs depending on their life history strategies.

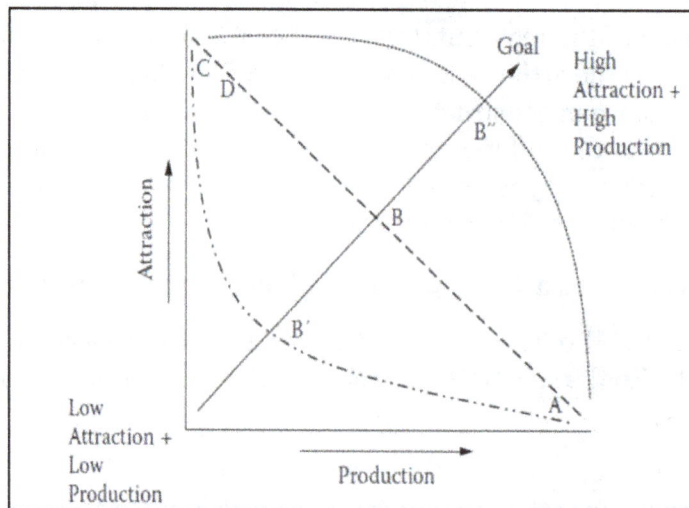

Figure: Relationship of A, B, C and D-type artificial reef species relative to attraction and production features of artificial reefs. B' and B" indicate the position of B-type species with different life history strategies

To attract A-type organisms, the artificial reef structures do not need to extend vertically into the water column but should be provided with internal spaces matching with the size of the target species, while for B-type fish species the holes should be larger and the artificial reef structures must reach at least a height of 2 m. To aggregate C-type species, the artificial reef should extend vertically into the water column and its structures should have wide open spaces to favour the water flow. Simple units can be also used for particular species, e.g. clay jars for octopus. Consequently, the complexity and diversity of the fish assemblage associated to an artificial reef strictly depends on the complexity of the reef.

Siting

The displacement of reef structures within an artificial reef may affect its influence on fish. Greater distance between the reef units/reef sets may increase the total volume of the artificial reef but account should be taken of the fact that the effects on fish may be reduced if the units are placed too far from each other. The reef groups may act as isolated reefs (in space) if spaced too far apart. Correct spacing of reef units (or modules), reef sets and reef groups will ensure that the reef complex operates as one reef with fish readily transiting between refuge/predation points, which can greatly improve the "productive" value of the overall reef complex itself. The area of influence (or halo) of each unit, set and group will vary depending on target species and module design.

In general, the criterion to be applied in positioning artificial reef structures within a reef group is that the areas of influence of individual reef units and/or reef sets should overlap with each other. The reef groups do not need to interact with each other when included within a reef complex.

Production artificial reefs should be placed in areas where stocks of target species already exist. Moreover, the reefs should match with the ecological requirement of those species. Usually in the Mediterranean Sea, this type of artificial reefs is placed in coastal waters up to 30 m depth, but the range depth can be appreciably greater in other seas (e.g. off Japan) where high relief artificial reefs are placed up to 80 m depth.

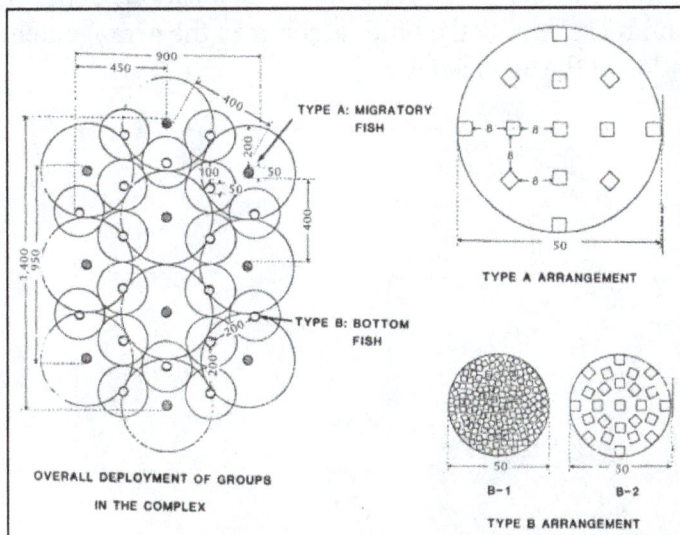

Figure: Spatial arrangement of reef units/reef sets in a reef complex

In the case of production artificial reefs aimed at enhancing and managing local fisheries, shifting the fishing effort helps compensate for the loss of fishing grounds due to other human activities.

The choice should be towards B-type species which are attracted to reefs to a limited degree but also gain some production benefit from reef platform. With respect to the above-mentioned criteria, to assure stability and ecological effects, the artificial reefs should be placed as close as possible to the fishing harbours, hence allowing equitable and economic access by reducing travel and search times, saving on fuel and increasing fishers' safety.

Artificial reefs may also help localize and manage the entire life cycle of some target fish. In this case, different reefs – each matching with the ecological requirements of a certain life stage of the species – should be deployed along the movement routes to gather the specimens in localized areas.

Practical Applications

France

Among the 94 000 m³ of artificial reefs existing in France, one third concerns the Marseille reef complex, which is the largest artificial reef deployed in the Mediterranean Sea with 27 300 m³ covering 220 ha and conceived by marine biologists. The reef deployment relied on the creation of horizontal and vertical discontinuities in heights, sizes and volumes thanks to a great variety of reef types and shapes, as well as on diverse arrangements and horizontal spacing of reef units/reef sets. Six types of modules of different shapes, sizes, volumes and materials were specially designed for this project. To optimize the reef habitat diversity, the complexity of these modules was enhanced adding several types of small filling materials (bags containing oyster shells, breeze blocks, octopus pots used for fisheries) and floating immersed ropes. Piles of quarry blocks of variable sizes were also used to reconstitute natural rocky boulders.

The different modules were grouped in six triangular shaped reef groups (300 m). These groups were linked together by series of reef structures (functional connections) functioning as biological corridors and stepping stones for fish and propagules. The locations of peripheral natural habitats (Posidonia meadows and rocks) were taken into account in the arrangement of the reef groups to favour a rapid colonization of the artificial reef.

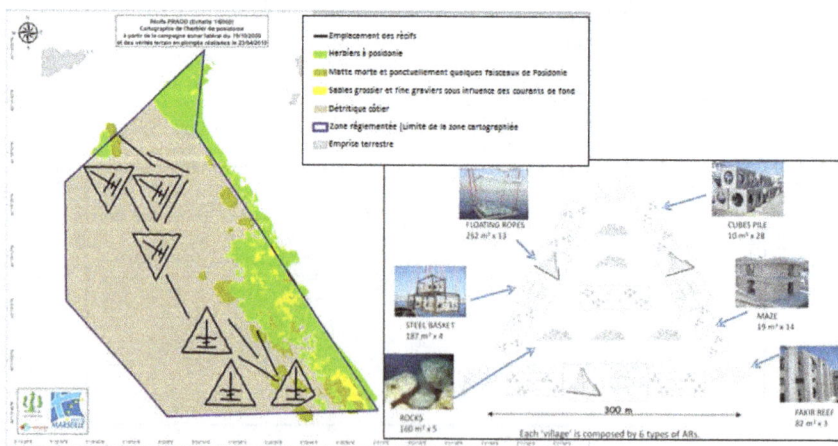

Figure: Marseille Prado artificial reef, the largest reef in the Mediterranean. It is composed of six "villages" linked by eight connections (above in green: lower limit of Posidonia meadow).
Each village has a triangular shape and is made with six types of artificial reefs

Greece

Four multipurpose artificial reefs were constructed in 2000–2006 for the protection and management of marine resources. The reefs have a surface area of 8–10 km² each and are made of different concrete modules. These are either mixed modules consisting of concrete cubic blocks with holes, deployed one by one on the seabed or assembled in pyramids, or production modules such as bulky cement-bricks on a concrete base and concrete pipes assembled in pyramids.

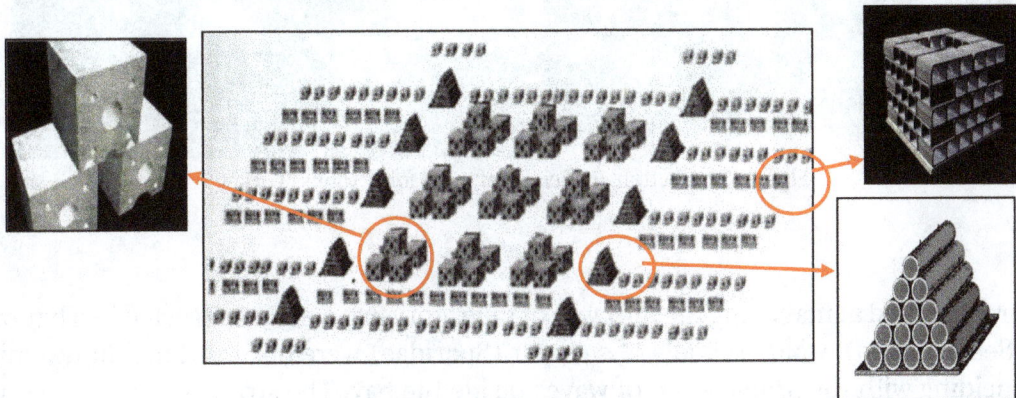

Figure: An artificial reef plan using four different types of modules in order to increase the reef complexity

Turkey

Octopus species are habitat-dependent and of great economic interest. Despite a lack of specific information, data show that the individual weight of octopus has decreased over the last few years both globally and in Turkish seas. Furthermore, natural shelters of Octopus vulgaris along the Aegean coast of Turkey are often disturbed by spearfishers. For these reasons, a plan was created to deploy an artificial reef specifically designed for this species (octo-reefs). The goal was to provide octopus individuals with suitable habitats and to increase their population in the long term. Simple concrete modules with holes were placed on the sea bottom. The first results showed that octo-reefs were actually used by octopus. Hence, the next step will be to deploy this type of artificial reefs in a closed fishing area and in MPAs.

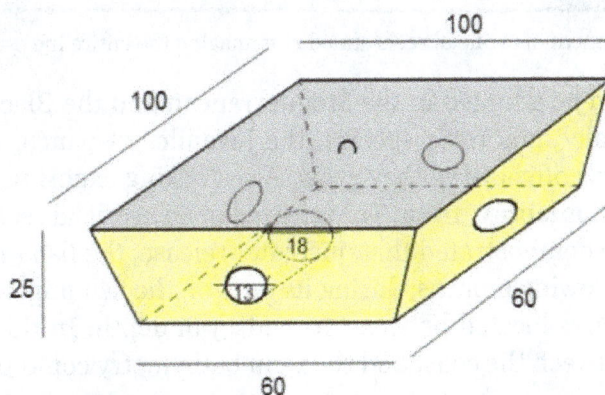

Figure: Production artificial reef units for octopus

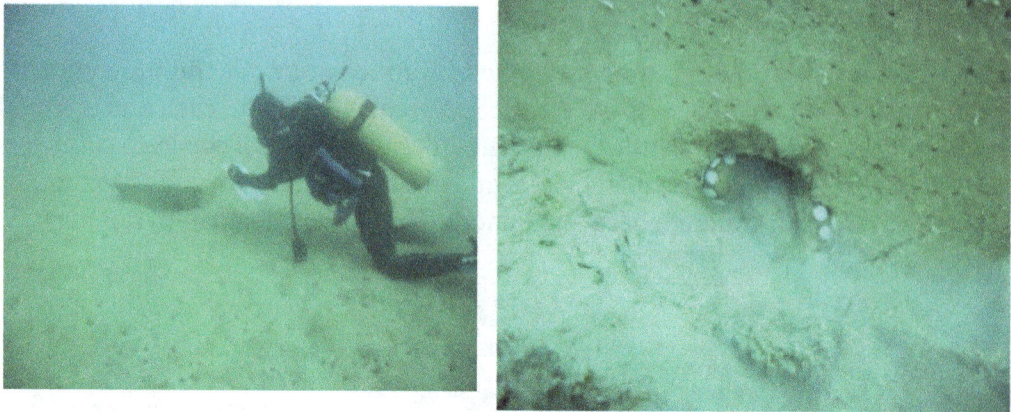

Figure: Production artificial reef units for octopus, Turkey

Japan

Artificial reefs aimed at managing the life cycle of migratory fish were constructed in a bay of the Iki Islands (Sea of Japan), where schools of snapper (Sparidae) were observed to follow a migratory route coinciding with the propagation of waves inside the bay. The strategy adopted was to place a production artificial reef at the entrance of the bay, a spawning reef where the waves converged, and a nursery reef to improve the survival of juveniles. This confined the life cycle of snapper into the bay, which considerably improved their survival, and allowed their catches to be managed by the local fishing communities.

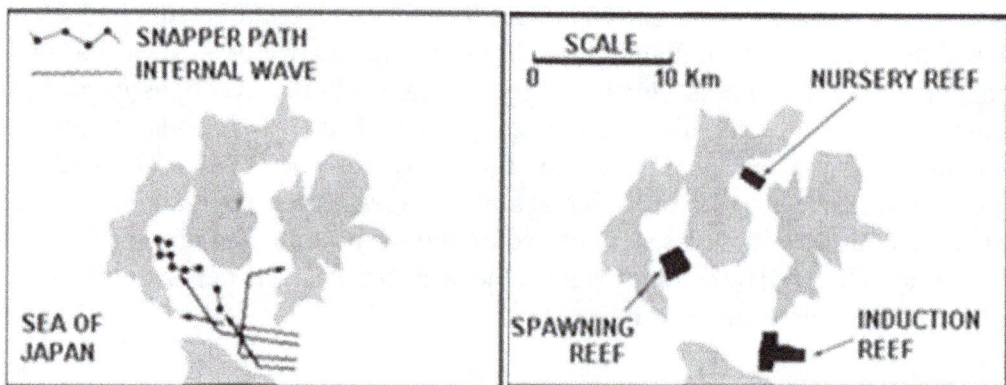

Figure: Deployment of artificial reefs aimed at managing the entire life cycle of snapper

Similar applications could be adopted in the Mediterranean and the Black Sea to manage the life cycle of some commercially important species, the juveniles of which, for example, prefer low depth and migrate towards offshore as they grow. A restocking experiment conducted with juveniles seabass (Dicentrarchus labrax; 15 cm TL) released in an artificial reef located at 11 m depth in the northern Adriatic Sea demonstrated that, just after release, the fish migrated inshore, close to estuarine areas. In the following months, during its growth, the fish migrated again to the artificial reef and the mussel cultures located between 10 and 13 m depth. In this case, the placement of suitable artificial reefs between the coast and the 13 m bathymetry could partially confine released seabass.

Open Ocean

Open ocean aquaculture is defined as the 'rearing of marine organisms under controlled conditions in the EEZ—from the three mile territorial limit of the coast to two hundred miles offshore. Facilities may be floating (for example, net pens for rearing of finfish and rafts from which strings of mollusks are suspended), submerged (fully enclosed net pens or cages moored beneath the water surface), or attached to fixed structures'. The terms 'open ocean aquaculture' and 'offshore aquaculture' are interchangeable.

The offshore process is founded on two rationales. The first; a need to redress the US seafood trade deficit, surpassed only by importation of oil and automobiles and currently running at just over $7 billion. The second; avoiding user conflicts over coastal aquaculture in locations that, 'are being increasingly contested by competing interests such as the fishing industries, recreational boaters, adjacent landowners and environmentalists'. More recently, self-sufficiency in food production has been offered as a solution to homeland security concerns.

Advantages of Offshore

Offshore fish farming advantages over inshore or onshore farming were stated as:

- Better fish flesh quality because of the action of waves and tide on the growing fish.
- Reduced diseases and use of antibiotics because of a more 'natural' marine environment.
- A smaller environmental impact due to a changing column of water around the cages.
- The economic advantages of scale which could be huge compared to inshore and onshore farming, because of the available space at sea.

Issues of Aquaculture

Aquaculture production in the offshore marine environment is not without its challenges. Fish are raised in an open system that can have serious consequences for ocean ecosystems if they are not sited, designed, and managed properly. There have been numerous examples of poor production practices that have impacted the surrounding environment by exposure to external stressors, such as: disease, chemicals, and therapeutants; excess nutrients that can impact biodiversity in benthic habitats; the release of non-native species that compete with native wild species; and potentially fatal interactions with wild species that are entangled in gear or intentionally killed as a perceived threat to the farmed stock. The perceived pressure put on wild stocks to use fishery products (fish meal and fish oil) to feed farmed fish, coupled with water quality and benthic effects from fish food and feces is another concern. While these are all relevant concerns, research has shown that proper siting and husbandry practices, best management practices, and the use of appropriate technologies and tools result in greater productivity while greatly minimizing and even eliminating some of these stressors altogether. The U.S. has the tools to develop a comprehensive regulatory framework and the resources to conduct proper oversight of the best management practices and ensure the use of the appropriate tools required for environmentally-responsible domestic marine aquaculture production.

Enhanced Stocking

The term "stock enhancement" is often broadly used to describe most forms of stocking, irrespective of purpose. From a fisheries view point, this can be somewhat misleading, even though the ultimate goal of every enhancement practice is to increase stock size and thereby, the fishable stock. Welcomme and Bartley recognize four major types of stocking intervention based on the objective of the intervention:

- Compensation - to mitigate a disturbance to the environment from human activities;

- Maintenance - to compensate for recruitment overfishing;

- Enhancement - to maintain fisheries productivity of a waterbody at the highest possible level; and

- Conservation - to retain or replenish stocks of a species that is threatened or vulnerable.

"Enhancements" are separated into two types:

- Stock enhancement of wild fisheries: The enhancement of stocks of an existing wild, open-access fishery with species that may or may not be self-recruiting. This category includes the stocking of relatively large inland waterbodies where there are no property rights to the stock. Generally the recapture rate of stocked fish is low and repeated enhancement is not always necessary to maintain the fishery.

- Culture-based fisheries: The stocking of small waterbodies is a form of enhancement that is typically undertaken on a regular basis and the stocking activity is the only means of sustaining the fishery. Typically, a person or a group of persons and/or an organization will have property rights to the stock. The source of stock for the enhancement may be derived from capture, but more typically is obtained from a hatchery operation. These features collectively amount to a form of aquaculture that according to the FAO definition, is referred to as culture-based fishery.

Apart from the above differences, the objectives for stock enhancement may differ markedly between developed and developing countries. Welcomme characterized the differing strategies with regard to management of inland waters for fish production, and these are equally applicable, with minor modification, to stock enhancement in inland waters The primary purpose of stock enhancement of floodplains, large reservoirs and lakes in Asia is to increase the foodfish supplies and is in contrast to that in developed countries, where it is to enhance recreational fisheries and for conservation purposes.

Stock enhancement in developing countries may be based on one of four broad strategies, or combinations thereof:

- Use as a seeding mechanism for replenishing depleted "breeding stocks" : e.g. the case of Indonesian reservoirs particularly using species that are indigenous to the country, but not necessarily to a particular waterbody .

- Replacement of existing, self-recruiting species/stocks, with species/stocks with more desirable traits (such as higher growth rate, reduced tendency to stunt etc.) e.g. the endeavours to replace Oreochromis mossambicus with O. niloticus, a practice that has become increasingly popular over the last three to four decades in most Asian lakes and reservoirs. More recently, there have been attempts to replace the original stocks of O. niloticus with the "GIFT" strain of O. niloticus (GIFT-genetically improved farmed tilapia).

- Regular stocking of species with a view to sustaining a fishery : In most instances, such stocked species are unlikely to form breeding populations in the waterbodies concerned. This is due to the stocked species requiring migration to a riverine habitat for breeding. Typical species used for this purpose include the Chinese and Indian major carps.

- Regular stocking of floodplains : as a form of compensation for reduction in recruitment resulting from developments related to flood control (e.g. in Bangladesh) and to increase fish yields, and/or develop new fisheries to enhance fish supplies (e.g. in Myanmar).

Table: Differing strategies for management of inland waters for fish production through stock enhancement

	Developed countries	Developing countries
Main objectives	Conservation	Provision of food
	Recreation	Employment
		Political
Mechanisms	Sports fisheries	Food fisheries
	Habitat restoration	Enhancement through intensive stocking (and management of eco-systems?)
	Environmentally sound stocking	Extensive/semi-intensive (+ integrated) rural aquaculture; culture-based fisheries
	Intensive, discrete, industrialized aquacultur	
Economic	Capital intensive	Labour intensive

Stock Enhancement of Inland Waters in Asia

Rivers and Floodplains

The enhancement of riverine stocks for fisheries development in Asia is relatively rare compared with developed countries. Stocking programmes for the Mekong giant fish species - the giant barb (Catlocarpio siamensis), the giant catfish (Pangasianodon gigas), isok barb (Probarbus jullieni), thicklip barb (P. labeamajor) and thinlip barb (P. labeaminor) are some of the few instances of riverine stock enhancement in Asia. Stocking of these species, some of which are endangered, is planned and/or in progress as a component of an integrated management strategy for improving the status of wild stocks.

In developed countries, there are few remaining artisanal freshwater river or floodplain fisheries. Riverine stock enhancement in these countries is carried out primarily for sport fishery development and for the conservation of indigenous stocks. A secondary purpose in some rivers is stocking for the control of aquatic weeds.

Riverine stocking with exotic species for the purpose of developing recreational fisheries, often associated with the promotion of tourism, has taken place in some Asian countries. These exotic species are typically salmonids, and the countries where this has taken place include India, Pakistan and Sri Lanka. This activity is still continuing to some degree, despite its potential negative affects on indigenous flora and fauna. Some enhancements have had negative impacts on native flora and fauna, such as the introduction of brown trout (Salmo trutta) and rainbow trout (Oncorhynchus mykiss) into New Zealand in the late 1800s, which is purported to have negatively impacted the native galaxiid stocks.

It is disheartening to note that the riverine stocking of exotic species has not been objectively evaluated and has attracted very little attention from the scientific community in the region. Indeed, attempts to justify culturing such exotics in the mountainous areas in the region have generally ignored the availability of local species that are equally or even better suited for this purpose (e.g. some species of Tor are excellent sportfish) and the increasing body of evidence of negative impacts on native fauna.

Stock enhancement of the giant river prawn, Macrobrachium rosenbergii, has been attempted in some rivers, large waterbodies and reservoirs in Thailand over a fairly long period of time; however, reliable data on stocking are available only since 1998. This is one of the relatively uncommon examples of stock enhancement with a non-finfish species. During the period 1998-2003, 15 Thai rivers were stocked in one or more years, with nearly 70 million postlarvae. The most intensely stocked river was Pak Panang in southern Thailand, which was stocked with 26 million postlarvae in 1999. Songkhla Lake has also been repeatedly stocked (in 2002, 11 million tiger prawn postlarvae, 7 million banana prawn postlarvae and 14 million giant river prawn postlarvae were stocked). Regrettably, however, there are few statistics available on the returns from these stock enhancement attempts. Although Choonhapran report some increased production following the stocking activities, no evaluation as to whether this was a direct result of the stocking activity could be made. Stock enhancements for giant river prawn have also been conducted in reservoirs.

Floodplains are wetlands that retain an association with the parental river and are typically inundated for part of the year during annual floods. These wetlands are very productive ecosystems and also provide crucial habitats for the spawning of some riverine species. The inundated parts of floodplains also provide important feeding grounds for fry and fingerlings of most riverine species. The floodplains of Bangladesh, Cambodia, Myanmar and several other Asian countries support substantial artisanal fisheries. In Bangladesh and Myanmar, some of these floodplain fisheries have stock enhancement strategies. In some cases, parts of the floodplain have been cut off from the parental river by damming for fishery enhancement and management. These fisheries are obviously managed in a more intensive manner and are more akin to "culture-based fisheries".

Lakes and Reservoirs

There are relatively few natural lakes in the Asian region, and the emphasis of stock enhancements has been mostly directed at reservoir stocking. Most of the natural lakes are not stocked regularly, and the available evidence indicates that stocking has been confined to self-recruiting exotic species such as tilapias (Oreochromis spp.) and common carp (Cyprinus carpio). In some countries such as Thailand, there may be an increasing trend towards the stocking of indigenous species capable of forming self-sustaining populations in large waterbodies. Details on such introductions and stock enhancements in Indonesian lakes are given in table. It is evident from table that in most instances, translocated indigenous species were not successful in establishing self-recruiting populations. In contrast, introduced exotic species such as Mozambique tilapia (O. mossambicus) and common carp were able to establish self-recruiting populations in almost all the lakes and have subsequently become the dominant species in their respective fisheries.

Due to their limited numbers, lakes are less significant in terms of fish production (except Tonle Sap Lake in Cambodia) and are not regularly stocked or enhanced. The stock enhancement of reservoirs is a major management strategy adopted for increasing fish production in these man-made waterbodies.

Table: Stock enhancement/introductions into Indonesian Lakes

Region/Lake	Size (ha)	Species (Year introduced)	Re-stocked	Comments
Sumatera				
L. Toba	112 000	Cyprinus carpio(1905)	na1	Contributes about 2%
		Oreochromis mossambicus (1940s)	na	Dominant in the fishery
Sulawesi				
L. Tondano	5 600	Trichopterus trichopterus (1925)		Yields of 340 kg/ha/yr; T. trichopteruscontributed about 10% to the yield but declined after the latter introductions
		C. carpio	na	
		O. mossambicus		
L. Limboto	7 000	O. mossambicus (1944)		30% of the yield (330 kg/ha/yr) in 1985-1991
L. Lindu	3 500	O. mossambicus (1950s)	na	Yields of 120 kg/ha/yr; contributes about 75-80% of the yield
L. Tempe	10 000 to 30 000	Trichogaster pectoralis (1937)	1940	Dominated the fishery until about 1948 Established
		Clarias batrachus(1939) H. temmincki(1925)		H. temminckidominated the fishery for a few years but declined rapidly with the introduction and repeated stocking of B. gonionotus, which accounted for most of the production (900 kg/ha/yr); but since 1982 yields are about 600 kg/ha/yr

		Barbonymus goniono-tus (1937)	Repeated since 1937	
Irian Jaya				
L. Sentani	9 360	Osphronemus gour-amy(1937)	1958	Apart from O. mossambi-cus, other species have not established self-recruiting populations and do not con-tribute significantly to the fishery, which yields about 42 kg/ha/yr
		T. pectoralis(1937)	1951	
		H. temmincki(1937)	na	
		C. carpio (1937)	1957	
		O. mossambicus(1951)	O. mossambicus(1951)	
		B. gonionotus(1966)	na	
L. Ayamaru	2 200	O. gouramy(1937)	na	All introduced and/or trans-located species, except O. gouramyare established in the Lake and C. carpiois the dominant species in the fishery (60%)
		T. pectoralis(1937)	na	
		H. temmincki (1937)	na	
		C. carpio(1957)	na	

The range of sizes of Asian reservoirs that are used for fishery activities requires that stock enhancement strategies of large (>600 ha), medium (<600 to >100 ha) and small (<100 ha) reservoirs are best considered separately. This is justified for the following reasons:

- The fisheries in large reservoirs in the region are usually "open access" and are not typically dependent upon stocking. These fisheries are based upon self-recruiting populations that may be "seeded" by stocking at irregular intervals.

- In small and medium reservoirs, the fisheries are almost always dependent on stocking, as natural recruitment is too small to sustain a fishery, even on a very small scale, and fishing pressure on the stocks is typically high.

- Fishery activities in small and medium reservoirs that are based entirely on a stocking and recapture strategy often also have well defined ownership. In such cases they are referred to as culture-based fisheries.

Seawater Ponds

In seawater pond mariculture, fish are raised in ponds which receive water from the sea. This has the benefit that the nutrition (e.g. microorganisms) present in the seawater can be used. This is a great advantage over traditional fish farms (e.g. sweet water farms) for which the farmers buy feed (which is expensive). Other advantages are that water purification plants may be planted in the ponds to eliminate the buildup of nitrogen, from fecal and other contamination. Also, the ponds can be left unprotected from natural predators, providing another kind of filtering.

Environmental Effects of Mariculture

Impacts of Mariculture on Water Quality

Effects of waste on water body depend on the key features of the habitat, waste quantity, time-scale over which the releases take place and characteristics of the water column. Aquaculture originated wastes can be subdivided into two major groups; solid waste and soluble waste. Solid wastes are composed of uneaten pellet feed, faeces, fish scale and mucus. Since most of them are denser than water, they sink to the bottom. Soluble wastes are ammonium, urea and some of the particles that solved from solid wastes. Among these wastes, nitrogen, phosphate groups and suspended substances are the main components that affect water quality. Moreover, some of the vitamins and minerals (trace elements) are effective on algae populations. Aquaculture originated pollution is generally divided into two categories as organic and chemical pollution.

Organic Pollution

The main sources of organic pollutants are fish faeces, uneaten feed and dead individuals. Faeces produced by fish are related with the feed ingredients. 25-50% of the consumed feeds are excreted to environment as faeces. Intensive fish farming causes large amounts of organic waste accumulation in the sediments and water column. A sediment collector placed under salmon cultivation cages collects 14.7-52 kg of organic waste in each square meter per annum, which reveals the extent of organic pollution originating from aquaculture activities. Aquaculture activities lead a number of ecological issues. For instance, aquaculture facilities such as cage systems significantly reduce water current speed thus cause accumulation of organic substances in surrounding environment. Decomposition of these organic wastes lead deterioration of water quality and increase water oxygen requirement. Negative effects like turbidity, accumulation of organic waste in the sediment, formation of anoxic zone near sea floor, accumulation of toxic substances and spread of diseases are influential in areas where aquaculture is in practice, especially in restricted exchange environments.

Formation of gasses such as H_2S (hydrogen sulfide), CH_4 (Methane) and CO_2 (Carbon dioxide) on sediment slows down growth rate of living organisms in the local area and makes them more vulnerable to diseases. Water oxygen level is reduced by direct decomposition of organic waste originated from aquaculture activities. Water quality changes by releasing nutrients like nitrogen and phosphate into the water column. Excessive amount of organic nutrients promote growth of water plants and algae. When toxin producing algae grow excessively, they may reach blooming concentrations and cause negative impacts to other organisms. Cultured fish may die off due to the disorders occurring in the gills and oxygen deficiency as a result of water quality changes related with sudden growth of algae. Consumers of bio toxin contaminated shellfish can get poisoned and die as well. When algae die in large numbers, dead organic substances start to decay and this process consumes oxygen, leads to hypoxia. Accumulation of organic and inorganic wastes in areas with low water exchange rate may lead hyper- nitrification as well.

In this sense, permanent aquaculture activities in such regions accelerate accumulation of waste. On the other hand, aquaculture activities held in deep waters or areas with high bottom current rate may result in wide spread of wastes. Let's make a comparison to explain extend of pollution caused by aquaculture activities. An average person excretes 4kg of nitrogen and 1.1kg of phosphate per year. For every ton of salmon 55kg of nitrogen and 4.8kg of phosphate are released into the water column through faeces per year. A salmon cultivation facility with a production capacity of50000 tons a year release an amount of nitrogen equals to that of 682000 people, and an amount of phosphate equals to that of 216000 people into the water in a year.

Eutrophication

Eutrophication simply means enhanced nourishment and refers to the stimulation of phytoplankton or macro fit (Algae and plants) growth in aquatic environment. To date, in all modeling studies, only effects of mariculture originated nutrient waste on phytoplankton growth were investigated. Especially, phytoplankton growth in summer months controlled by nutritious elements (Nitrogen, phosphate etc.) were modeled. Factors of this kind of model may lineup as; determination of bottom structure, water column affected from waste and dilution rate, calculation of nutrient level and determination the relationship between phytoplankton growth and nutrients.

In Baltic Sea, relations between sea water sensitivity, (depends on the interaction between open water and coastal water) nutrient concentration, phytoplankton biomass and secchi depth compared by using multiple regressions techniques. Depending on the sensitivity of the coastal areas, while phytoplankton biomass increases, secchi depth decreases. But since the water in shallow areas is under influence of wave- induced re suspension, secchi depth method is unreliable. Therefore, the most accurate method for determination of the phytoplankton biomass is the measurement of chlorophyll content. In another study, direct and indirect effect of eutrophication was investigated between 2001-2011 in Baltic Sea by using nutrient concentration, chlorophyll-a, and secchi depth indicators. Study revealed that entire open Baltic Sea was affected by eutrophication between 2007 and 2011.

Chemical Pollution

A variety of chemicals   used for disinfection and disease outbreak treatment in aquaculture. Environmental impacts of chemical compounds depend on toxic effect of the compound, pathogens sensitivity to antibiotics, therapeutics and active time zone of the chemical in the environment. Most of the chemicals were adapted from the poultry and cattle husbandry and their possible effect on the marine environment isn't fully investigated.

Chemicals in use in mariculture can be classified as disinfectants, anti foulings and veterinary medicines. Main veterinary medicines are antibiotics, anesthetics and pesticides. Veterinary medicines are widely used in fish farms but it is not so common in crustacean and bivalve cultures. Most of the chemical treatment scenarios in mariculture result in direct release of chemicals into water column and sediment. For the cage systems Atlantic salmon, rainbow trout, sea bass and sea bream are the most common fish species and most of the chemicals used for which cause pollution. Among numerous definitions of marine pollution the most common one is GESAMP's (Joint Group of Experts on the Scientific Aspects of Marine Pollution), which states that marine pollution is "the damage to living creatures in the environment, the hindrance to marine activities including fishery, the inability to use sea water and the reduction of amenities as a result of changes in environmental conditions due to deleterious effects induced by various waste materials and/or energy left to the marine environment (including estuaries) by people directly or indirectly.

When a disease emerges, farmers try to control the outbreak by using antibiotics. As a result of that, antibiotic makes its way through water by uneaten antibiotic treated feed or feaces. In order to control external parasites such as sea lice, different kinds of pesticides are used. Pesticides could be given to fish via feed or by bath. Pesticides such as ivermectin have high toxic effects on marine organisms. Bio fouling is one of the major problems in cage farming. Proliferation of bio fouling organisms adds considerable weight to open water cage net which threats the stability of the structure and readily reduce water flow through which reduces water quality within the cage systems. Anti-fouling paints and coatings are used in order to slow growth of such organisms on net to maintain water flow through to systems. Active matter of such paints is generally copper. Degradation of the paint in time causes copper to penetrate water, which poses toxic effects on marine creatures. Another toxic metal originated from aquaculture is zinc. Zinc sulfate ($ZnSO_4$) is used as feed additive to prevent disease outbreaks.

Antibiotics

Chemicals have wide range of usage in fish health management. Report prepared by GESAP and WHO (World Health Organization) mention chemical usage pose a threat to human health. Despite the reduction in antibiotic and organophosphate use in aquaculture, synthetic parathyroid, colorant, preservatives, anti-parasitic and other chemicals that cause marine pollution are still in use. Use of chemicals not only affects marine environment but also pose a risk to the farmers as well. Indiscriminate use of antibiotics for fish health management is one of the major problems in aquaculture. Unmetobolized antibiotics are often passed to the aquatic environment by uneaten antibiotic treated feed and by faeces. In both water column and sediment, different classes of antibiotics have different half-life. oxytetracycline which is one of the most common antibiotics can be effective up to 30 days in both water and sediment. In fact, oxytetracycline is still measurable on sediment after several months without being biologically active. In cage systems, chemicals in uneaten antibiotic treated feed and faeces scatter to the surrounding environment by waves and current. Wild fish and filter feeders such as mussels and oysters consume antibiotic treated feed or it passes to sediment.

Bioactive substances such as antibiotics and pesticides are generally used for controlling disease and parasite in aquaculture. Success or failure in aquaculture depends on the correct usage of chemicals, usage duration and usage conditions against infectious diseases and parasites. Effects of bioactive matters on environment can be align as;

i.   Substances with inhibitory effect last long in animal tissue.

ii. Release of substances with inhibitory effect to the aquatic environment.

iii. Development and transmission of the antibiotic resistance gene in microorganisms.

Substances with Inhibitory affects Last Long in Animal Tissue

Bioactive substances last long in living cells more than expected. For instance, trimethoprim in used in disease treatment, can be detectible in a mussel after 77 days. In order to totally eliminate oxytetracycline and other sulfonamide based antibiotics from trout body above 10°C, at least 60 days should pass after the last usage. Frequent use of antibiotics to prevent or treat bacterial fish diseases, may result environmental bacteria and intestinal bacteria to gain antibiotic resistance. Antibiotics make their way to muscles through digestion system. The emergence of antibiotic resistance bacteria as a result of massive use of chloramphenicol was determined in trout farms, in Italy. Antibiotic taken into body through digestion emits from intestines and passes to vascular system, from where is passes to muscles. Every antibiotic has a different withdrawal period. Since the muscle tissue of the fish carries antibiotics for a specific period of time after the treatment, fish should be kept in water for a while depending on the water temperature in order to eliminate the antibiotic residue from the tissues before putting fish on market. For instance, fish treated with oxytetracycline shouldn't be offered for sale at least for 20 days. Especially in summer months along with temperature rise, mortalities in market size fish force farmers to use antibiotics. But farmers generally don't heed withdrawal time procedures. These antibiotic containing fish offered to market before the withdrawal period pass to the consumer causing bacteria in human body gain antibiotic resistance.

Releasement of Substances with Inhibitory Effect to the Aquatic Environment

Frequent use of the inhibitor substances in aquaculture and replacement of potential bioactive substances to the aquatic environment is frightening. Only 20-30 percent of the antibiotic treated feed is eaten by fish, whereas 70-80 % of it passes directly to the aquatic environment. Oxytetracycline decay fast in marine environment but most of them bind to the tiny particles and accumulate on the sediment. Accumulated oxytetracycline could still exist at enough concentration to exhibit activity even after 12 weeks. Sediment that contains antibiotics effects the marine fauna negatively. For example, oxytetracycline residue was detected in blue mussel (Mytillus edulis) 80 meters away from cages where the antibiotic was used.

While the antibiotics on the sediment surface lose its effect easily, antibiotics like oxytetracycline, oxolinic acid, sarafloxacin and flumequin within the sediment can be effective for up to 180 days. Antibiotics on the sediment become ineffective so easily not because of the degradation, but due to the spread. Antibiotics like sulfadiazine and trimethoprim have relatively less half-life within the sediment which is approximately 90 days. In the other hand, half-life of the florphenicol in sediment was calculated as 4.5 days. Various natural and synthetic chemicals, such as dichlorvos, malachite green and derris root are used in large amounts in mariculture worldwide. We might give an example in order to reveal total amount of chemical usage; in 1989, Norwegian farmers used 3488 kg of dichlorvos in order to control sea lice (Lepeophtheirus salmonis) outbreak. Disclorvos and most of the used chemicals cause serious impacts on environment and they are need to be used wisely.

Development and Transmission of the Antibiotic Resistance Gene in Microorganisms

Antibiotics are widely used for controlling and treatment of both human and fish diseases. The amount of antibiotic use in aquaculture and in human health management is nearly same. Thus, frequent usage of antibiotic cause development of antibiotic resistance gene in both human and animal pathogens and treatment of the bacterial diseases with antibiotics is getting harder and harder each day. Antibiotics are not only using as disease controller but also as growth and feed conversion promoters. In 2000, Denmark banned all non-therapeutic uses of antibiotics. In 2006, EU banned the use of antibiotics as growth promoter. Antibiotics used at low doses increase the development of antibiotic resistance genes in human pathogens and opportunist environmental bacteria and ease the path of gaining multiple antibiotic resistance of bacteria. Plasmids (extra chromosomal elements) located in plasma of the bacteria carry resistance genes. Development of the antibiotic resistance gene arises from the changes in plasmid DNA and chromosomal DNA mutation.

Resistance plasmids (resistance, R) can be transferred between bacteria spp. Plasmids are transferred between bacteria in two different ways; Transduction and conjugation. Most of the bacteria species including pathogenic bacteria have plasmids. Transfer of R Plasmids from resistant bacteria to sensitive bacteria cause different bacteria species develop antibiotic resistance gene. Being transferable and independently replicateable plasmids are the main factor of wide spread of the resistance genes. Since the excessive use of antibiotics in aquaculture could promote development of antibiotic resistance genes in pathogens, most of the antibiotic treatments are inefficient. Roughly 16 different antibiotics are commonly in use in aquaculture. Ampicillin, florphenicol, erythromycin neomycin, oxytetracycline and tetracycline are among them. Some of the bacteria gain resistance against these antibiotics inevitably. In 1981, resistance of Aeromonas salmonicida the etiological agent of the furunculusis disease to oxitetracycline and was 4%, resistance ratio reached 50% in 1990. Presence of antibiotic resistance gene in the environment reduces effectiveness of the antibiotic. Usage of the antibiotics as a growth promoter, leads development of antibiotic resistance in human pathogen bacteria and increases the risk of antibiotic resistance gene transfer. Transfer of the R Plasmids from aquatic microorganisms to human pathogens poses a potential threat to human health.

Impacts of Mariculture on Sediment

To date, the importance given to investigation of impacts of pesticides, antibiotics and synthetic paints on human health has not been given to the studies related to investigate the impacts of these chemicals on marine environment. After feeding fish in a cage, uneaten feed and faeces accumulate on the sea floor. Excessive accumulation of organic substances on sea floor leads excessive proliferation of microorganisms which causes reduction in oxygen level required for other benthic organisms (crustacean, fish etc.).

Disease outbreaks occur rarely in wild fish populations since they are scattered all over the oceans and seas, whereas high population density in aquaculture systems often leads to disease outbreaks. Accumulation of uneaten feed, faeces, chemicals and antibiotics causes formation of anoxic zone on sea floor thus, most of the benthic organisms that live near to the salmon farms dissappeared, only some of the resistant bacteria proliferated. In many countries, formation of antibiotic resistance in poultry, cattle and aquaculture borne bacteria threats human health. Antibiotics used against bacterial fish disease outbreaks are generally excreted to the water column. These antibiotics accumulate on sediment or animals living in surrounding environment. Scientists detected 9 different antibiotic compounds in different stream sediments. The concentration of the antibiotics ranged from 1.01µg/kg (sulfachloropyridazine) to 485µg/kg (sulfamethoxazole). Under suitable environmental conditions, antibiotic residues on sediment effect microbial communities and only resistant individuals can survive.

Impacts of the Pollution on Microbial Communities

In marine ecosystem, heterotrophic bacteria play an important role in biodegradation Bacteria are very sensitive to the environmental changes thus they can be affected directly from anthropogenic nutrient input. Studies clearly show that, the pollutions caused by aquaculture activities are changing structure and activity of benthic communities. Chemical medicines are one of the most influential factors on bacteria communities. Direct release of antibiotic treated feed and medicines to the marine environment cause selective pressure on bacteria populations, fish pathogens and leads to development of antibiotic resistance genes.

Most commonly and widely used antibiotics groups in aquaculture are oxytetracycline, florphenicol and sulfonamide. Oxytetracycline taken by feed cannot be metabolized 100% by digestion. Some active quantity is excreted after metabolism and finds their way to water by faeces. Antibiotics have a significant impact on the microbial populations that live on sediment. Biomass of the bacteria and balance of different species of bacteria is changing by antibiotic use. Bacteria living in sediment are responsible for nitrogen, phosphate and sulfate cycles and create the first step of food chain. Antibiotic contamination reduces the formation of sulfate and nitrate. Effects of this outcome on microbial communities are not clear. Another issue which needs to be known is, which compounds antibiotics degrade to after metabolization and what are the potential harmful effects of these chemicals. For instance, when florphenicol is taken by salmon, it transforms into florphenicol amine in fish body. To date, impacts of the metabolites substances on marine organisms have not been fully studied. The most studied topics are related with the development of the antibiotic resistance gene due to the accumulation of antibiotic residues in surrounding areas of the aquaculture facilities.

Pollution Originated from Construction/Building Materials

Some of the construction materials expose heavy metals, plastic substances and by-products to the marine environment. Presence of these substances is generally unknown to farmers but we become aware of them day by day. The usages of preservatives, which are believed to be harmless, are gradually increasing. Plastic substances contain fatty acid salts, chromates, cadmium sulfide, antioxidants, UV conservatives, organophosphate, fungusites and disinfectants. Most of these composites are toxic for aquatic life but due to their low solubility and slower decay rate, their toxicity decreases.

Impact of toxic substances that leaks from construction materials of aquaculture facilities on environment are not fully understood. Docks, bottom of the vessels and nets painted with anti foulings contain effective and inexpensive chemical called tributyltin which is active substance of most of the antifouling. Water pollution caused by tributyltin inhibits shell formation of oysters, kills mussel larvae, causes imposex, and endocrine disruption, in molluscs and also contaminates sediment, sea water, fish and other bivalve spp. Therefore, usage of substances that contains tributyltin is a big problem in some of the countries. Fortunately, worldwide use of thetributyltin as a antifouling agent has been banned by The International Maritime Organization (IMO) in 2003, new antifouling agents have been introduced to the market as a replacement.

Indirect Effect of Pollution on Human Health

Bacteria can adapt to the changing environment with presence of extra chromosomal elements (Plasmid) and horizontal gene transfers of the plasmids. Adaptation could be gained by realignment, duplication, by changing the copy number of the plasmids or by the lateral and horizontal movements of the plasmids between bacteria populations. Plasmids affect life functions of the host since the plasmids have vast range of bacteria host in nature and ability to multiply within the host. Resistance genes that developed on plasmids cause changes in genetic character of the bacteria and can be transferred both vertically (transfer of the genes between organism in a manner of reproduction) and horizontally (transfer of the genes between organism in a manner other than reproduction). In natural microbial communities genetic adaptations arise from presence of a plasmid via horizontal gene transfer. Transfer ratio of the plasmids in polluted environment is 2 to 10 times more than thatin unpolluted environment. Plasmids play an important role in development and spread of the antibiotic resistance genes.

References

- Shellfish-aquaculture/intertidal-system, shellfish-farming-101: bcsga.ca, Retrieved 14 July 2018

- Pubmed-17094702: ncbi.nlm.nih.gov, Retrieved 17 May 2018

- Shellfish-Farming: americanmussel.com, Retrieved 11 June 2018

- Offshore-fish-farming-needs-a-social-licence-to-keep-growing: worldfishing.net, Retrieved 30 June 2018

- Aquaculture-Workshop-WEB: aquariumofpacific.org, Retrieved 14 May 2018

Chapter 5

Environmental Issues

Unregulated aquaculture practices can be environmentally damaging. Some of the concerns are the adverse effects of antibiotics, improper waste handling, farmed and wild species competition, invasive plant and animal species introduction, etc. This is an important chapter, which analyzes the environmental impact of aquaculture such as pollution, habitat destruction, disease transfer, etc.

Environment Impact of Aquaculture

Negative Impacts of Aquaculture on the Environment

In South-East Asia where finfish and shellfish are heavily produced and poorly managed there are fairly heavy environmental impacts. Finfish production here is usually quite intensive and involves an addition of solids and nutrients to the marine environment to help fish grow. This process is generally recognised as being potentially degrading to the environment as such a rapid unnatural build-up of organic material can negatively impact on the localised flora and fauna. In some cases this can cause major changes to the sediment chemistry and affect the overlying water column location.

The effect of farmed fish on local wild fisheries is also a real environmental concern in South-East Asia and elsewhere. Outbreaks of disease from farms can spread quickly due to the high concentrations in which fish are retained and is easily spreadable into wild fish populations if uncontrolled. Aquaculturalists used to tackle these outbreaks with antibiotics in fish feed until concern mounted over the effect of the drugs on local aquatic ecosystems as well as on consumers. Vaccinations are however now readily available for farmed fish and the practice of using drugs to tackle disease is seldom used in Western aquaculture.

Additional impacts related to aquaculture may also occur as a result of other farm discharges and waste products. These can include from shore-based stun and bleed operations, the escaping of non-resident species, transmission of disease and (lack of) control of predatory species. Where species such as shellfish compete with other organisms such as native seagrass for survival, displacement can occur which has a potentially spiralling effect on the native wildlife.

Positive Impacts of Aquaculture on the Environment

Despite a negative outlook there are some fairly positive environmental impacts to be recognised from aquaculture. These can be found in (artificially or naturally) nutrient enriched areas where the farming of filter feeders such as shellfish improve water quality. Farmed fish are also generally free of environmental contaminants such as mercury and heavy metals as they exclusively eat human-processed feed of which toxin levels are regulated.

Pollution

The relation between aquaculture and pollution is a complex one. Animals living in a water body are affected by changes in the chemical and physical quality of that water. An increase in water temperature can be beneficial up to certain point, above which it becomes detrimental and ultimately lethal; but a change involving an increase in level of a chemical which is not normally present can very often cause deleterious effects.

The animals are sensitive to the physical and chemical changes to which many of them have a low tolerance range. Even changes within their tolerance range can affect the physiology of the animals so as to influence their growth rate, fecundity and mortality. The tolerance of the animals also varies with their stage of development. Thus a very young animal may be more sensitive to a pollutant than an adult. This account is intended to bring together the information available on the common pollutants and their tolerance levels in the cultivable fishes, crustaceans and molluscs that are important in aquaculture.

Nature of the Culture Systems and their Relation to Problem of Pollution

Fishes living in a culture system are subjected to changes in water to a greater extent than those in the natural surroundings as they cannot select the environments in which they live. The fishes and prawns living in an estuary or lagoon can often escape from a local pollutant by moving to other areas. Obviously this is not possible for fish confined in a pond or cage.

Based on the mode of feeding of the culture animals, two types of aquaculture systems can be distinguished. In the first one, fishes depend for food on the plants and animals produced in the farm area in which they live as against the second one in which the fishes are confined in dense populations in small areas such as cages and small ponds and wholly depend on artificial feeds. In both these systems the sensitivity to changes in water quality differs. In the first system pollution affects not only the culture animals but also the animals and plants on which they feed. This system, therefore, is likely to be more easily damaged by pollutants than the second one where the damage inflicted Is directly on the culture animals.

Biological Effects of Pollutants

Of the many different toxic substances that find their way into the estuaries and coastal waters and affect the physiology of the animals inhabiting them, pesticides occupy the top most places. The agricultural sector has using different types of organochlorine and organophosphate compounds for the eradication of pests. Recent studies on the effects of insecticides on marine organisms demonstrate that concentrations which are not sufficient to control many species of insects, nevertheless, can inhibit, the productivity of phytoplankton; kill or immobilise crustaceans, fishes and molluscs; kill eggs and larvae of bivalve molluscs; induce deleterious changes in tissue composition of molluscs and teleosts; affect distribution of schooling and feeding behaviour of fishes; and interfere with ovary developments in molluscs and teleosts Boyd, 1964 Eisler, 1970. It has been observed that clams and oysters can concentrate pesticides from the medium by factors of 70,000 and more. Fishes also can concentrate appreciable quantities of insecticides directly from the medium and retain them for at least 4 months. Marine species are unable to acquire resistance to pesticides and suffer heavy mortality when exposed to relatively low pesticide levels. In general organochlorine insecticides are more toxic to marine fishes and crustaceans than organic phosphate insecticides and detergents.

Criteria for Water Quality in Fish Culture

Domestic garbage and industrial effluents containing materials that precipitate in seawater, settle on the bottom or float, can cause trouble to aquaculture. Shellfish may be killed if the beds of the farms are covered with settling substances. Materials from mining operations also can cause the same deleterious effects to shellfish by causing disturbance to their feeding mechanisms. Industrial effluents, depending upon their intensity and nature, can cause havoc to both culture animals as well as the plankton forms on which they feed. Measurements conducted with C have indicated that the effluents discharged into Chaliar river (near Calicut, Kerala) affect photosynthesis right up to the Beypore estuary, 16 km downstream from the point of impact.

The use of garbage, dairy waste and sewage as organic fertiliser to increase productivity of fish ponds has been practised for a long time. Organic wastes such as cow-d jng and manure from poultry and piggery can be directly used for fertilising fish ponds; but in the case of sewage, it is to be subjected to primary and secondary treatments with high rate trickling filter or by activated sledge process.

Before recycling domestic waste water through fish pond, it is desirable to treat it so that the organic load is reduced considerably. The organic load, of waste waters is generally expressed in terms of its Biochemical Oxygen Demand (BOD) which in the case of raw sewage generally lies between 150-600 mg/1. Primary treated sewage effluent contains less organic matter than raw sewage and more nutrients than secondarily treated one and is preferred for fish culture in ponds where no supplementary fertilisers of foods are used.

Beneficial effects in water quality from domestic and animal wastes, industrial organic wastes and heated effluents require careful study by aquaculturists so that a pollutant which in effect is a "displaced resource" can be profitably exploited for producing much required protein for filling up the nutrition gap. Coupled with this, bio-assay studies are also required to monitor the uptake of unwanted or toxic substances in the lipid pool by bio-accumulation in order to maintain the quality and safety of products of aquaculture.

Habitat Destruction

If aquaculture is operated indiscriminately, environmental damage is often the consequence, especially in coastal areas. This can occur with mussel farming or fish farming in cages, where there is direct contact between the aquatic animals and the surrounding waters. In the past, farmed fish such as European Atlantic salmon in North America often escaped from their cages. In time they transferred diseases to the wild populations on the US coast. If the alien species feel at home in their new environment they can breed prolifically and in some cases completely crowd out indigenous species. Cultivation of the Pacific oyster was abandoned some decades ago by mussel farmers in Holland and off the North Sea island of Sylt. The species has become a problem, spreading over the entire mudflat area of the Wadden Sea – a shallow coastal sea bordering the North Sea – and overrunning the blue mussel, the staple food of the eider duck and the oystercatcher. The banks of mussels have now become inaccessible to the birds. "Invasive alien species" is the term given by experts to these non-native species. Regulations in Europe now govern the introduction of new species, prescribing a long period of quarantine. Many areas of Asia, however, do not take the problem of invasive alien species nearly so seriously. For this reason experts are calling for in-depth case-by-case assessments of the potential of species becoming prevalent in a new habitat and changing the ecosystem. Another problem can be the removal of juvenile fish or fish larvae from their natural habitat. The European eel, for example, migrates from the rivers of Europe to spawn in the Sargasso Sea in the western Atlantic. As this species cannot be bred in captivity, juvenile eels must be caught in the wild for breeding purposes. The practice places extra pressure on wild eel stocks. Happily, however, increased public pressure has virtually put a stop to mangrove clearances for new fish farms in the major river estuaries of South East Asia. The mangroves also proved to be unsuitable for the industry. Like the Wadden Sea mudflats, the sediment in mangrove forests contains nitrogen compounds, in particular toxic hydrogen sulphide. For several reasons this environment proved to be inappropriate for farming. According to development aid agencies, aquaculture facilities based on brackish water are no longer being established in the mangroves in Thailand, but in areas further inland.

Escapes

When fish or shrimp are kept in pens or ponds that are connected to natural waterbodies, some can escape. This isn't as harmless as you might think. In some cases, escapees can impact wild populations by competing with them for food, habitat and spawning partners.

Most farmed species are distinctly different than their wild cousins. If escapees breed with their wild counterparts, the genetic makeup of their offspring may be less suited to surviving and thriving in the wild. In addition, fish are sometimes farmed in areas they are not native to. If they escape, they can establish themselves as invasive species and disrupt the harmony of the ecosystem.

Preventing Fish Farm Escape

Preventing farmed fish from escaping and interacting with wild fish is a priority for:

- Governments
- Indigenous peoples
- Environmental groups
- The aquaculture industry
- Commercial and recreational fishers.

To minimize chances of escapes, finfish containment systems (such as net-pens) must be able to withstand local weather and ocean conditions. These conditions include storms, water currents and other environmental factors. The containment systems must be regularly inspected and maintained to ensure integrity and to control factors that could contribute to failures, including:

- Biological matter buildup
- Ice buildup
- Marine mammal interactions.

If breaches happen, licence holders must report escapes to the responsible regulatory authority. We may also approve fishing to recapture escapees, where it's warranted and effective.

Disease Transfer

Contaminants exist in wild and farmed seafood. The nature of contamination depends upon the species, geographic region, animal age and diet, and production practices. Among farmed seafood, the main contaminants of concern are methylmercury, persistent organic pollutants (POPs), and production drugs. Farmed fish acquire POPs and mercury from consuming fish meal and oil, which are feed products made from small, wild pelagic fish such as herring and sardines. POPs and mercury in the ocean come from atmospheric deposition of emissions, mainly from the combustion of fossil fuels for mercury and agricultural pesticide application for POPs POPs and mercury in coastal waters come from discharge/runoff from industrial sources and from contaminated land and sediments Farmed salmon tends to have lower levels of POPs and mercury when fish-based feed is sourced from regions of the world where small pelagic fish have fewer contaminants, such as South America, compared with small pelagic fish from Northern Europe, which have higher levels of contaminants Some fish processors are implementing activated carbon filters to remove POPs from fish meal and oil as a means of reducing contaminants in farmed salmon Another approach to reducing POPs in farmed fish is to reduce the use of fish meal and oil in aquaculture.

Among wild-caught fish, methylmercury, heavy metals, and POPs are the most concerning contaminants with regard to public health. These contaminants are primarily found in apex predatory fish such as shark or swordfish, and in bottom feeders such as tilefish. These fish acquire contaminants as a result of bio-magnification up the food web, and from decades of intensive industrial development in the northern hemisphere. Consuming these fish is not advised by national governments and epidemiologic studies have quantified the risks from consuming these fish, particularly among vulnerable groups such as pregnant women, in populations that consume large amounts of seafood Levels of mercury, other heavy metals, and POPs in wild-caught fish are generally lower in the southern hemisphere, but are increasing.

Scientists have attempted to assess the trade-offs between the health risks associated with heavy metals and POPs in seafood and the benefits of omega-3 LCPUFAs One study reported that high levels of mercury can mask the beneficial cardiovascular effects of DHA Another review article concluded that the cardiovascular benefits of regular intake of farmed salmon with high levels of omega-3 LCPUFAs outweigh the theoretical risks associated with chronic exposure to moderate levels of polychlorinated biphenyls (PCBs) The findings of Roth and Harris align with a previous systematic review showing that for adults, the benefits of fish consumption outweigh the risks In women of childbearing age, the benefits of seafood intake also outweigh the risks, if some species high in mercury and POPs are not consumed In a recent systematic review, positive associations were found between prenatal or postnatal seafood intake and child neurodevelopment in most studies; however, the counterbalancing effects of pollutants (POPs and methylmercury) and health benefits can make it challenging to observe health benefits Risk communication about seafood consumption is evolving as the science on health effects evolves. Unfortunately, both consumers and the media have been confused by public health messaging about the benefits and risks of seafood, which may suppress consumption among groups such as pregnant women, who could benefit from eating non-contaminated seafood high in omega-3 LCPUFAs.

Aquaculture and Production Drugs

A variety of chemicals (algaecides, antibiotics, disinfectants, herbicides, pesticides, and probiotics) are used in aquaculture to treat and prevent diseases, in an attempt to achieve maximum production Contamination of farmed seafood products and the environment due to the use of these chemicals can cause risks to human health, potentially increasing the prevalence of some NCDs such as cancer Antibiotics used in medicated feed can diffuse into the water column, spreading to sediments and wild fauna In laboratory trials, some antibiotics have remained stable and retained antimicrobial activity for months in marine sediments and the insecticide diflubenzuron has remained stable for at least 7 months in sediment Research has documented an increase in antibiotic-resistant bacteria associated with antibiotic use in aquaculture The USA, European Union, Chile, and many other countries inspect only a fraction of commercial products for environmental chemicals, microorganisms, pesticides, and veterinary drugs to safeguard the food supply. Given the human health risks associated with the use of chemicals and veterinary drugs in aquaculture, this problem fits well into a One Health framework. To address this issue, these drugs should be used minimally, and veterinary drugs should be used under the supervision of a veterinarian and regulated by the appropriate government agency. In addition, investment in research aimed at understanding the underlying causes and epidemiology of animal diseases is critical for ultimately reducing the use of production drugs.

Social Impacts and Inequity

Industrial-scale aquaculture products are largely exported to other countries or sold to middle- or high-income individuals in the country of production, resulting in situations in which people living near aquaculture production are impoverished and food insecure. This is especially true in the case of populations that previously used the land/water for subsistence fishing, as is true for many sites now used for industrial shrimp farming in Southeast Asia Environmental impacts from aquaculture can affect the well-being of coastal communities by changing residents' sense of place, decreasing community involvement, and compromising mental health In addition, aquaculture's use of fish meal and oil competes with the availability of fish for human consumption The dependency on fish for food is higher on islands and in coastal communities, especially in low-income countries, where fish and other seafood represent a high proportion of dietary animal protein. For these populations, small fish are the main source of micronutrients that help to combat dietary deficiencies Alternatively, aquaculture can help people living in poverty by providing employment and sustaining local economies; this may be especially true for women, who constitute much of the post-harvest aquaculture workforce Therefore, it is essential to examine aquaculture operations on the basis of their impact on living conditions and how/if they meet the needs of consumers at various income levels, both proximal to the production site and for export markets.

Environmental Challenges: Aquaculture and Environmental Health

Maintaining a high degree of environmental quality is critical to aquaculture, because of multiple ecologic feedback loops linking human health and seafood production. Aquaculture may affect human health and nutrition by reducing wild fish populations (because of its use in aquaculture feed) or by causing environmental impacts and spreading fish diseases that reduce future aquaculture or fisheries production. Scarcity of wild fish and/or aquaculture products would not only impact food security but could also increase consumption of foods that promote the development of NCDs.

A significant challenge facing aquaculture is increasing seafood production to the levels needed to positively impact diets at a population level without degrading aquatic ecosystems. In addition to overfishing concerns related to feed ingredients, there are impacts on the local environment at many aquaculture sites, associated with the chemicals used on farms, effluent discharges and water quality, disease transmission between farmed and wild species, concentration of fish waste, and fish escapes.

Production systems that interface directly with the environment and rely on ecosystem services, such as clean water and fish waste decomposition, have a greater likelihood of negatively impacting their surroundings, compared with recirculating systems that are sited on land. This is perhaps why policies and siting of new offshore aquaculture operations are particularly contentious. Therefore, a better understanding is needed regarding the trade-offs between different aquaculture production methods such as offshore and recirculating systems.

Environmental Challenges: Aquaculture and Animal Health

Each year, aquaculture producers lose large amounts of farm-raised seafood because of infectious disease outbreaks, which cost billions of dollars, impact international trade, and generate negative publicity for the aquaculture industry. Over the past 30 years in Southeast Asia and South America, over a dozen emerging viral diseases have spread throughout shrimp farms, some causing very high mortality rates. Over the past few years, infectious salmon anemia (ISA), a viral disease, has impacted Chile by causing significant reductions in salmon production and exports. Although most aquaculture diseases do not directly impact human health because the pathogens do not infect humans, the chemicals and drugs used to prevent or treat them can impact the environment and public health.

A variety of factors contribute to emerging diseases in aquaculture, including:

   (i)   Globalization and international trade;

   (ii)   Consolidation and intensification of hatcheries and production;

   (iii)   Introduction of hatchery-raised species to new environments;

   (iv)   Interactions between wild and farmed animals;

   (v)   Biosecurity; and

   (vi)   Climate change.

Specific risk factors have been developed for particular species, regions, production methods, and diseases. In Chilean salmon culture, risk factors for infectious salmon anemia were related to insufficient surveillance and diagnostic efforts, poor management practices, close proximity of farms, high prevalence of sea lice (Caligus rogercresseyi), insufficient disease prevention and contingency plans, poor stock quality control, and insufficient transportation practices. Additional issues include potentially suboptimal feed formulation and a scarcity of local research and development in fish nutrition. In Southeast Asian shrimp production, risk factors for white spot syndrome virus include high stocking densities, use of wild stocks raised in ponds, use of alternative and live feeds, poor water quality, and animal stress.

Given the complex set of interactions that facilitate the spread of disease, multi-level interventions are necessary. Farm-level disease interventions, such as timely diagnosis and treatment, could address the host–pathogen relationship, while environmental stewardship and improved farm management may address environment–pathogen and environment–host issues. To address disease transmission between farms, regional and national policies, surveillance, reporting, training, and emergency response capabilities are also needed.

References

- The-impact-of-aquaculture-on-marine-habitats, eco-friendly-aquaculture: worldoceanreview.com, Retrieved 15 July 2018

- Environmental-impacts-of-aquaculture: greentumble.com, Retrieved 20 May 2018

- Escapes, aquaculture, ocean-issues: seafoodwatch.org, Retrieved 25 April 2018

- Escape-prevention-evasions-eng, protect-protégé, aquaculture: dfo-mpo.gc.ca, Retrieved 10 June 2018

Chapter 6

Aquaponics

Any system that integrates aquaculture and hydroponics in a symbiotic environment is known as aquaponics. In this system, the water from an aquaculture system is sent to a hydroponic system and the by-products are broken into nitrites and nitrates by nitrifying bacteria which are then used by plants. The filtered water then recirculates back to the aquaculture. All such important aspects of hydroponic subsystem, deep water culture, fish stocking, etc. have been covered in this chapter.

Aquaponics is a combination of aquaculture, which is growing fish and other aquatic animals, and hydroponics which is growing plants without soil. Aquaponics uses these two in a symbiotic combination in which plants are fed the aquatic animals' discharge or waste. In return, the vegetables clean the water that goes back to the fish. Along with the fish and their waste, microbes play an important role to the nutrition of the plants. These beneficial bacteria gather in the spaces between the roots of the plant and converts the fish waste and the solids into substances the plants can use to grow. The result is a perfect collaboration between aquaculture and gardening.

Aquaponics is a big hope for sustainable organic crop production, aquaculture and water consumption. The fish waste is recycled and used for plant growth instead of throwing it in the ocean. The water is recirculated in a closed system lowering the consumption of this resource.

Types of Systems

Since aquaponics uses basically the same systems as hydroponics, there aren't many differences in how the system works, except for the added fish in the water tank(s). Drip irrigation, flood and drain, deep culture or water submerged roots, and nutrient film technique are highly compatible and customizable to merge with growing fish.

Importance of Ph Control in Aquaponics

pH is an important part of aquaculture. Setting it to a perfect level can be a bit confusing since there are three living organisms to care for: your plants, your fish, and the bacteria inside the water and each of them has a different pH need. A neutral pH from 6.8 to 7.2 is good for the aquaponic garden. Because of the fish waste, the pH will become acidic and you will need to use aquaponic compatible pH adjusters. If the pH level is not beneficial for the system that is too low or too high, the plants will not be able to absorb nutrients optimally and your fish will die eventually. It's very important to monitor the pH level each day and to keep it within the neutral range.

A too alkaline or too acidic pH is one of the main reasons fish or plants die, leading to gardening failure. The pH adjusters need to be specially designed for this type of growing system, otherwise, they could harm the fish. You can find these adjusters in a local aquaponic gardening supplier.

Another thing to keep in mind is the water hardness because it affects how pH will behave when trying to adjust it. Sometimes it would be necessary to also take care of the water hardness when working the pH. Fish don't like sudden changes in pH, so when adjusting it try to lower or increase it slowly.

Fish and other Aquatic Animals you can Grow in Aquaponics

Fish are the ones feeding your plants. The fish used in this type of aquaculture are freshwater fish, most popular being tilapia and barramundi because they tolerate better diverse water conditions and they grow fast. Trout can also be used especially for lower water temperatures. Other aquatic animals you can grow are snails and shrimps.

You can feed the fish special food you can purchase in an animal store or other foods like water lettuce and duckweed.

Benefits of Aquaponics

1. Aquaponics is a way to grow your own fish and vegetables at the same time. You feed the fish and the fish will feed your plants through their waste output.

2. There is no need to use fertilizers because the fish provide rich nutrients for the plants.

3. In aquaponics, less water is used for the crops. Research has shown that aquaponic gardens use 1/10th of the water you would use for soil garden.

4. Regular gardening pesticides or other chemicals can't be used because they would harm the fish.

5. This results in healthier and organic vegetables.

6. You won't experience any soil borne diseases in aquaponics because there is no soil.

7. You can grow plants in very small space, and have a great harvest.

8. Plants grow fast because they get very nutritious substances from the fish waste.

9. Plants and fish production can be done in a controlled temperature environment.

10. Water is used in a closed system and circulated effectively, reducing the consumption and the water bills.

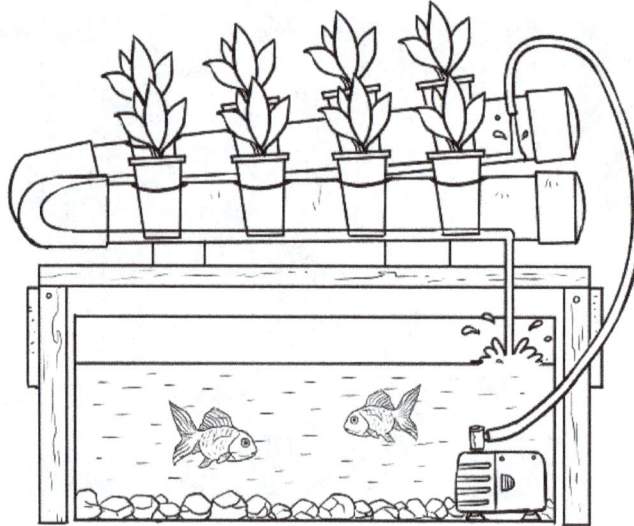

Tips on an Aquaponic Garden Set up

1. You can make your own aquaponics system and here is a simple and complete guide you can use to make one. Start small see if it's good for you then feel free to go bigger.

2. Have set a different power source as a backup. It's vital to keep the water flowing and the oxygen pumps on.

3. Make sure you feed the fish enough and let them thrive. Depletion of fish stock makes this type of cultivation impossible.

4. Keep food input constant for the fish and that will result in regular fish waste you can use to feed your plants.

5. Ensure your plants and fish with good aeration. Not only the plants need their roots to be oxygenated, but also, the fish and the bacteria need the water to be oxygenated. As the fish grow bigger, their oxygen needs increases and you might need to adjust accordingly.

6. When you decide what plants you want to grow, pick the ones that have similar water condition needs as the fish, and you will have greater success.

7. Remove some excess fish waste when necessary. Too much can harm the health of the fish.

8. Keep an eye on the level of pH because as shown above, it is crucial for the garden.

9. Fish tanks should be made of glass or food grade plastic.

10. Avoid using any pesticide other than organic, or any other substances that can and will harm the fish or the good bacteria (vinegar, citric and/or hydrochloric acid).

Components of an Aquaponic System

There are three main components of an aquaponics system: plants, fish, and bacteria.

Plants

The whole purpose of the aquaponics system is to grow plants in an environmentally sustainable way allowing for food security. The plants do not just receive all the benefits of the aquaponics system. They actually play a highly important role in maintaining the overall cycle of the aquaponics system. They act as a natural filter for the water, absorbing the nitrates, therefore detoxifying the water allowing for recirculation back to the fish. The plants remove the need to clean waste accumulated in the fish tank because they use the waste, absorbing nutrients for growth, specifically nitrates, that can be toxic to the fish.

The plants sit on a grow bed, a container that holds the nitrate rich water, and floats the growth base. At ISB, we are using cement mixing containers as our grow bed. Tip: make sure the grow bed container is opaque, preferably black. A clear or translucent container creates a potential for algae growth because of the light entering the container. Opaque containers create a much lesser chance of algae growth because they block light from entering the areas they cover. A lightweight, buoyant, base is needed to hold the plants upright. Insulation foam blocks work excellently because they are easy to cut holes in order to place the plants, and they are sturdy enough to hold their place. The grow base can be cut in order to just have room for the plants, or it can be cut to accommodate small net pots in order to allow for easier removal. It is necessary for the pots to be net style in order to allow maximum water exposure and ample room for root growth.

Figure: Grow bed containers (cement mixing containers)

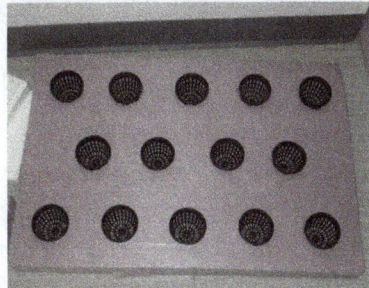

Figure: Pots placed inside the holes of the grow base

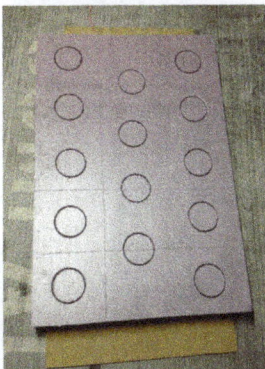

Figure: Insulation foam gridded, and holes scored

Figure: Net container/pot

Fish

This is the aquaculture component of the aquaponics system. The excretions from the fish are what eventually provide the nutrients needed to grow the plants. For every pound of fish, about two gallons of water are needed; for every gallon of water, there can be The fish tank is the area that needs to be the most maintained, just because they are living animals, which also means that they are one of the main indicators of overall health of the system.

Several factors have to be considered when selecting fish, especially because they are going to be living in a tank environment. In order to have maximum growth output, the fish needs to be able to live in high-density population conditions. Also rapid growth is optimal considering the growth of the plants is dependent on the waste excretion from the fish. The fish species also has to be able to withstand living in an enclosed space (that is, the tank).

Tilapia is normally used because of these very standards. They can thrive in a tank environment, they grow rapidly, and they are a high-density living fish. Other common aquaponics fish are perch, catfish, trout, and hybrid striped bass.

The one input (other than replenishing water because of evaporation) after the initial set-up of the aquaponics system, is the fish food. Depending on the fish species, feeding standards will differ. All the fish food has to be the type that is able to float on the surface.

Bacteria

From first glance, it may seem that there are only two components of an aquaponics system: fish and plants. However, there is one other piece that without it, there would not be an aquaponics system. The bacteria play a highly integral part of the cycle. The bacteria are in between the fish and the plant stage. The bacteria are what transform the waste into the nutrients able to be absorbed by the plants. They do this through a process known as nitrification.

Nitrification is the process in which nitrogenous organic compounds are converted into nitrites, then nitrates. The first step is the conversion of ammonia to nitrite. This is done by Nitrosomas (in soil) and Nitrosoccus (in aquatic environments). The ammonia is oxidized by these bacteria into nitrites, which then flow to the second group of bacteria. Nitrobacter (in soil) and Nitrococcus (in aquatic environments) further oxidize the nitrites into nitrates. Once converted into nitrates, the compounds are in a form that can be absorbed by the plants.

These bacteria can be present in a biofilter. This is located in between the clarifier and the grow beds. It can also just be present in the grow beds and/or the fish tank.

Secondary Components of the Aquaponics System

- Aerator: constantly aerates the water allowing for more oxygen to enter the promoting better fish health, and more rapid plant growth.

- Pipes: pipes (usually pvc) are what transport the water in between the various stages of the cycle.

- Lights: the lights provide the radiant energy needed for plants to photosynthesize.

- Pump: the pump is the main electrical source that pushes the water.

Here is a summative, simple, figure that provides an overall visual of the main components of an aquaponics system.

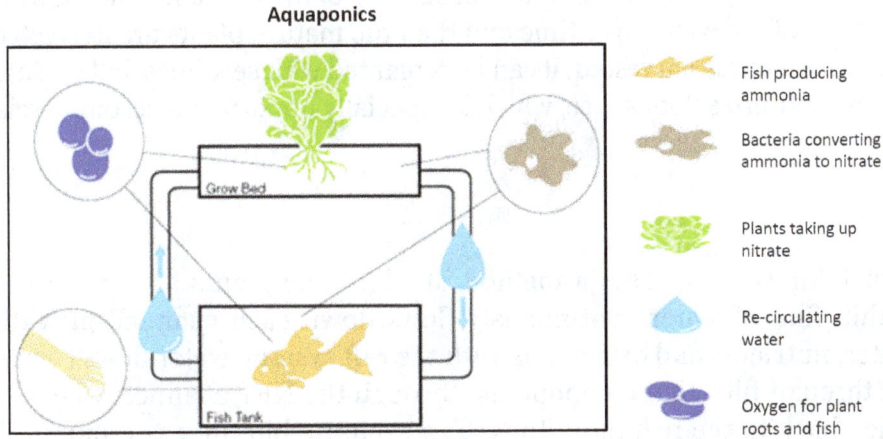

Aquaponics

Hydroponic Subsystem

There are three primary aquaponic methods emerging in the industry; raft, NFT and media-filled beds. Each of these methods is based on a hydroponic system design, with accommodations for fish and filtration.

Raft

In a raft system (also known as float, deep channel and deep flow) the plants are grown on polystyrene boards (rafts) that float on top of water. Most often, this is in a tank separate from the fish tank. Water flows continuously from the fish tank, through filtration components, through the raft tank where the plants are grown and then back to the fish tank.

The beneficial bacteria live in the raft tank and throughout the system. The extra volume of water in the raft tank provides a buffer for the fish, reducing stress and potential water quality problems.

This is one of the greatest benefits of the raft system. In addition, the University of the Virgin Islands and other research programs refined this method during 25 years of research.

In a commercial system, the raft tanks can cover large areas, best utilizing the floor space in a greenhouse. Plant seedlings are transplanted on one end of the raft tank. The rafts are pushed forward on the surface of the water over time and then the mature plants are harvested at the other end of the raft. Once a raft is harvested, it can be replanted with seedlings and set into place on the opposite end. The optimizes floor space, which is especially important in a commercial greenhouse setting.

NFT

NFT (Nutrient Film Technique) is a method in which the plants are grown in long narrow channels. A thin film of water continuously flows down each channel, providing the plant roots with water, nutrients and oxygen. As with the raft system, water flows continuously from the fish tank, through filtration components, through the NFT channels where the plants are grown and then back to the fish tank. In NFT, a separate bio filter is required, however, because there is not a large amount of water or surface for the beneficial bacteria to live. In addition, the plumbing used in a hydroponic NFT system is usually not large enough to be used in aquaponics because the organic nature of the system and "living" water will cause clogging of small pipes and tubes. NFT aquaponics shows potential but, at this time, it is used less than the other two methods.

Media-filled Bed

A media-filled bed system uses a tank or container that is filled with gravel, perlite or another media for the plant bed. This bed is periodically flooded with water from the fish tank. The water then drains back to the fish tank. All waste, including the solids, is broken down within the plant bed. Sometimes worms are added to the gravel-filled plant bed to enhance the break-down of the waste. This method uses the fewest components and no additional filtration, making it simple to operate. The production is, however, much lower than the two methods described above. The media-filled bed is often used for hobby applications where maximizing production is not a goal.

Deep Water Culture

Deep water culture is a type of hydroponic system in which the plant's roots are submerged in a growth-inducing mixture containing essential nutrients and minerals. In this system the plants are aerated via an air pump.

In early aquaponics, the deep water culture system included an air pump, air stone, and a net pot that contained the grow medium along with a five gallon bucket.

Some plants, such as lettuce, thrive in water and are commonly grown using deep water culture.

To build a deep water culture system, horticulturalists recommend the following equipment: grow media, cups, pots and baskets, air stones, air line, aquarium air pump, and a container that will hold the nutrient solution.

There are various ways to provide aeration. The most popular uses air bubbles, with air stones or even an aquarium air pump to diffuse air bubbles into the water's nutrient solution. Air stones are porous, and these pores create individual pockets of air.

A different variation of circulating air is through falling water. While it is not the most common option, falling water creates surface agitation that aerates the nutrient solution. Higher falling water creates more agitation, which consequently creates more dissolved oxygen.

Biofilter

Biological filters are devices to culture microorganisms that will perform a given task for us. Different types of organisms will perform different tasks. Part of the art of designing and using biofilters is to create an environment that will promote the growth of the organisms that are needed.

Importance of Biological Filters for Aquaculture

We use biofilters to help maintain water quality in recirculating or closed loop systems. Biofilters are also used to improve water quality before water is discharged from a facility. There are many different methods of maintaining good water quality and biofiltration is only one component of the total picture. It is however, a very important and essential component especially for recirculating aquaculture or aquarium systems.

Need for Biofilters

Depending on design and application, biofilters have the ability to accomplish the following functions. The first three functions are performed by biological means and the last four are done by physical processes that do not depend on living organisms.

1. Remove ammonia

2. Remove nitrites

3. Remove dissolved organic solids

4. Add oxygen

5. Remove carbon dioxide

6. Remove excess nitrogen and other dissolved gasses

7. Remove suspended solids.

In general, there are two types of aerobic microorganisms that colonize biofilters for aquaculture. Heterotrophic bacteria utilize the dissolved carbonaceous material as their food source. Chemotrophic bacteria such as Nitrosomonos sp. bacteria utilize ammonia as a food source and produce nitrite as a waste product. Chemotrophic bacteria such as Nitrospira sp. utilize nitrite as a food source and produce nitrates as a waste product. Nitrosomonos and Nitrospira will both grow and colonize the biofilter as long as there is a food source available. Unfortunately, both of these types of bacteria are relatively slow growing. Heterotrophic bacteria grow about 5 times faster and will out compete the other two types for space if food is available. Since most aquaculture biofiltration systems are designed for the purpose of converting and removing ammonia from the water this presents a problem.

There are three ways to deal with this problem. The first is to remove most of the carbonaceous BOD (biological oxygen demand) before the water enters the biofilter. The second method is to provide sufficient extra capacity (surface area) in the biofilter to allow all of the various bacteria to grow. Another method is to have a very long plug flow path through the biofilter. This allows different zones of bacteria to establish themselves in different parts of the biofilter.

There are 4 main types of aerobic biological filters and several subcategories of each. Here is a listing of the major types.

1   Recirculated Suspended Solids (Activated sludge and biofloc systems)

2   Aquatic plant filters

   • Unicellular (Microscopic)

   • Multicellular (Macroscopic)

3. Fluidized Bed Filters

- Sand Filters

- Bead Filters

4. Fixed film

- Rotating Biological Contactors (RBC)

- Trickling Filters

- Submerged Filters (with or without aeration)

    a. Up flow

    b. Down flow

    c. Horizontal flow

    d. Moving Bed

Anaerobic filters can also be defined as biofilters but they are never the main biofilter used for maintaining water quality in the culture system. There are two main reasons why they are not suitable. The number one reason is that they are not capable of effectively cleaning the water to the level required. The other reason is that they operate too slowly. Anaerobic filters are sometimes used in aquaculture for conversion of nitrates into N2. However, this is a difficult process to control and it is generally less expensive to remove nitrates by discharging a small amount of water from the system. The water removed with the solids is usually sufficient to remove the nitrates as well.

Anaerobic biofilters are best suited for processing high strength waste. The sludge produced by the physical filter system is an example of a high strength waste. Processing plant wastes are another candidate for anaerobic digestion. In an integrated production/processing plant these two streams could be combined. The best feature of anaerobic systems is the production of methane. There are specially designed engines that can burn this gas to produce electricity. Using the gas to heat water is another obvious possibility. However, the capital cost of these systems generally limits their use to large operations.

Characteristics of the Ideal Biofilter

Before we examine each type of biofilter, it would be useful to define the characteristics of the ideal biofilter. The following characteristics can be considered a checklist that we can use to rate each of the different types. In some cases, different features may be mutually exclusive but we can use the ideal characteristics as a yardstick or goal. In practice it may be necessary to trade off one feature for another but it doesn't hurt to know what the ideal should look like. The following list contains most of the pertinent features of a good biofilter.

1. Small footprint - The biofilter should occupy as little space as possible. It is common to have culture tanks and the biofilters under cover for protection and temperature control. Space allocated for biofilters takes away area that could be used for culture tanks.

2. Inert materials of construction - All materials used in the biofilters should be non-corrodible, UV resistant, resistant to rot or decay and generally impervious to chemical attack. In general, marine grade construction materials are required for reasonable working lifetimes.

3. Low capital cost - The biofilter must be inexpensive to purchase or build and cheap to transport to the farm location.

4. Good mechanical strength - The biofilter and its components must be tough enough to withstand the normal wear and tear of an industrial/agricultural environment.

5. Low energy consumption - The energy cost (usually electricity) to operate the biofilters should be as low as possible. The largest energy users are the pumps to move water and compressors to move air.

6. Low maintenance requirements - The biofilters should be self cleaning with little or no care required for the normal life of the crop.

7. Portability - The biofilters should be easily movable to facilitate changes in operation of the facility.

8. Reliability - Ideally the biofilters should have no moving parts that could fail at an inopportune time. If the biofilters does have moving parts, they should be rugged and designed for a continuous operating life of several years.

9. Monitorabilty - It should be easy to observe the operation of the biofilter to insure that it is operating correctly.

10. Controllability - It should be easy to change operating variables to assure optimum performance.

11. Turndown ratio - The biofilters should be able to work under a wide range of water flow rates and nutrient loading levels.

12. Safety - The biofilters should not have any inherent dangers to either the crop or the owner/operator.

13. Utility - The biofilters should accomplish all of the goals set forth in beginning of this paper i.e. removal of ammonia, carbon dioxide, BOD, suspended solids etc.

14. Scalable - A small system should work the same way as a large system. The performance per unit volume should be constant regardless of the size of the system.

Now that the characteristics of the "ideal" biofiltration packing have been established, it makes sense to compare the existing medias to that standard.

Characteristics of Real Biofilters

Activated Sludge Systems

Activated sludge systems are not common in aquaculture systems. Activated sludge systems are good at removing carbonaceous BOD in systems with high nutrient loadings. They are commonly used in domestic waste water treatment systems. Activated sludge systems are typically expensive to operate and do not provide the effluent water quality necessary for aquaculture.

Aquatic Plant Systems

Plants are not normally used for the primary biofilter in aquaculture systems. They do however provide a very good sink for the nitrates produced by a well functioning biofiltration system. The marriage of recirculating aquaculture systems and hydroponics is known as aquaponics. Aquaponics use the feed resources efficiently and effectively. In addition to the valuable plants grown in aquaponic systems, they minimize the amount of waste that must be disposed of in the environment. Removal of nitrates and phosphorus from waste water is a big benefit.

Unicellular plants (algae, diatoms etc.) are sometimes allowed to grow in the culture tanks. Some species such as tilapia are tolerant of poor water quality and can use the algae as food. Systems operated this way are sometimes called "green water" or biofloc systems to distinguish them from the clear water systems that many species require. Green water systems can be a very cost effective way to culture certain species but they are not recommended for beginners to aquaculture. Management of these systems requires some experience and specific knowledge.

Fluidized Bed Sand Filters

Regular sand filters such as the type used for swimming pool filters or potable water filters are virtually worthless as biofilters for aquaculture. The biofilm quickly fills the spaces between the grains of sand and the pressure drop across the filter rises rapidly. Frequent back flushing is required and the active biological film is removed each time. In contrast, fluidized bed sand filters have been successfully used for aquaculture applications. A sand filter becomes fluidized when the velocity of the water flowing up through the bed is sufficient to raise the grains of sand up and separate each grain from its neighbors. In hydraulic terms, the drag on each particle is sufficient to overcome the weight of the particle and the particle is suspended in the stream of water. The velocity required to fluidize the particle is a function of the shape, size and density of the particle.

Fluidized bed sand filters have several very good advantages. They pack more biologically active surface area into a given volume than any other type of biofilter. In addition, the best shape for a fluidized bed sand filters is a tall column. Thus they have a small foot print for a given capacity. They are self cleaning and relatively tolerant of different nutrient loadings.

There are also several disadvantages and potential problem areas with fluidized bed sand filters. The fluidized bed sand filter has a relatively high energy requirement because of the high pressure drop necessary to fluidize the sand. The other main problem with sand filters is that the pressure required to fluidize the bed varies depending on the amount of biofilm on the sand particle. As the biofilm builds on the sand particle the size of the particle increases while the density of the particle decreases. This means that the depth of the bed will tend to increase as the bed ages. It also means that the bed depth will fluctuate as the loading on the bed varies. In order to prevent blowing the sand out of the tank, the tank must be oversized or the flow of water needs to be regulated.

Another potential problem is the uniformity of the water flow. In order to completely fluidize the bed, the water needs to be evenly distributed across the whole bed. Two things can happen if the flow is not uniform. One possibility is that the water will channel and short circuit though the bed. This means that the treatment capacity will plummet. Another possibility is that the short circuit will happen near the wall of the vessel and the abrasive sand will eat a hole through the wall of the vessel.

Fluidized bed sand filters are limited to the oxygen carried in with the water. This means that the water entering the filter should have a high level of oxygen in order to insure a good level of treatment.

Bead Filters

Bead filters are a relatively new type of biofilter. They are advertised as the complete solution to water quality for recirculating systems. They consist of a closed vessel partially filled with small beads of plastic. Usually the vessel is filled with water and the beads float at the top of the vessel. Water flows up through the bed of beads. The beads are small enough to trap most large suspended solids. In addition, the surface of the beads supports the growth of a biofilm. The small size of the beads means that they have a relatively large surface area per unit volume. The larger systems incorporate a mechanical stirring device such as a propeller on a shaft. Periodically the water flow is shut off and the bed of beads is agitated to dislodge the suspended solids. The solids are allowed to settle into the bottom of the vessel and then drained off. This ability to remove suspended solids and act as a biological filter is advertized as the main advantage to bead filters.

The difficulty in successfully operating bead filters lies in striking a balance between the competing functions. Too frequent washing to remove solids dislodges the biofilm and disrupts the nitrification process. If the beads are not washed enough however, the solids start to plug the bed. The other potential problem is the presence of large amounts of carbonaceous solids which tends to encourage the growth of heterotrophic bacteria at the expense of the autotrophic bacteria that work on the ammonia and nitrites.

Another drawback to bead filters is their relatively high energy consumption due to their high pressure drop. Also, the water flow and pressure drop are not constant. As the bed of beads becomes loaded with solids, the pressure drop rises and the water flow decreases. This leads to cyclic rather than constant performance.

Since bead filters are not aerated, they are limited to the oxygen carried in with the water. In general this is not a problem since retention times are low. Bead filter systems are probably suitable for small, lightly loaded systems where labor costs are low. At this time they are not available for large systems except as multiple units.

RBC (Rotating Biological Contactors)

Like much of the equipment used in aquaculture, RBC's were first used in domestic sewage treatment applications. There are several different types available for aquaculture. A typical design consists of plates or disks that are attached to a horizontal shaft. The shaft is located at the surface of the water and it is turned at a very slow speed (1-5 rpm). The disks are half submerged in the water at all times. As they rotate, the biofilm attached to the surface of the disk is alternately exposed to air and then submerged in the water. The original designs used an electric motor to turn the shaft. There is a new design specifically for aquaculture that uses compressed air or pumped water to drive a paddle wheel in the center of the cylinder. These RBC's float in the water and do not require bearings or elaborate mechanical supports.

RBC's have many advantages. They offer excellent treatment efficiencies. They require very little energy to operate and can be located in the culture tank to save space if necessary. They do not require additional oxygen and are not limited to oxygen contained in the incoming water. They can remove dissolved BOD or ammonia depending on nutrient levels. They are biologically robust and handle shock loads well. It is easy to observe their operation and visually monitor the biofilm. They only have one major drawback besides cost and that relates to reliability. If there is a power failure or the cylinder stops turning for any reason, the biofilm exposed to the air can dry out. When this happens, the cylinder will be unbalanced and can become difficult to turn.

Trickling Filters

Trickling filters are one of the oldest types of biological filters. Trickling filters filled with rock or coal were built in the late 1800's for sewage treatment. Trickling filters typically consist of a packing or media contained in a vessel. The water to be treated is sprayed over the top of the media and collected in a sump underneath the media. The surface of the media or packing provides the substrate for the growth of a biofilm. In large systems, air is forced into the filter with a fan. However, small can filters rely on natural convection and diffusion to move air throughout the filter.

Trickling filters are rugged and easy to operate. They have the ability to treat a wide variety of nutrient levels. Properly designed systems can handle solids very well. One of the big advantages of a trickling filter is that the water can leave with more oxygen than it entered. Because trickling filters have a large - air water interface, they also act as strippers to remove CO_2, H_2S, N_2 or other undesirable volatile gases. The only major drawback to trickling filters is the energy cost required to pump the water to the top of the filter. A high narrow filter will save space but take more pumping energy. A wide low filter will use less energy but take up more space.

The first step in the design of a trickling filter is to pick the right packing or media. Over the years many different materials have been used for trickling filters but for the last 40 years, the best packing has been structured media. Structured media is composed of sheets of rigid PVC that are corrugated and glued together to form blocks. For an in depth review and analysis of packing materials, refer to the paper "A Review of Biofiltration Packings".

One of the advantages of structured media is its flexibility and ease of use. Structured media can be used to build a small biofilter without a vessel. Since the vessel is typically the major cost of a biofilter, a biofilter with no vessel can be a real money saver. Structured media can be stacked on a frame work or any flat surface. It can be located over a culture tank or have its own water collecting sump. No sides are required because the packing is self supporting. Of course, large systems are typically built with walls and fans to move air through the media.

The most important requirement in the design of any trickling filter is a good water distribution system at the top. There are several ways to do this. A pressure spray system with splash guards at the top is probably the simplest. The only drawback is the additional pressure drop required to operate the nozzle. The other system involves the construction of a shallow water distribution pan with several gravity flow target nozzles in the bottom of the pan. Here are some typical arrangements for a "vessel-less" trickling filters.

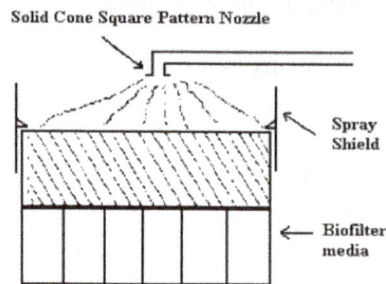

Figure: Trickling filter with pressure nozzle distribution system

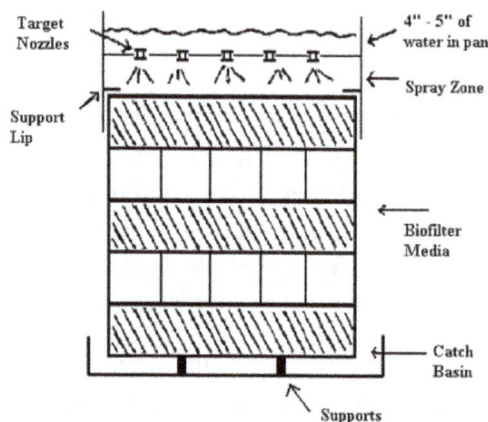

Figure: This is a trickling filter with gravity flow target nozzles in a shallow water distribution pan

Part of the art of designing a trickling filter is to balance the competing requirements on the design.

1. In order to keep the energy costs to a minimum, the pumping head for the filter should be as low as possible. The maximum plan area covered by the filter is determined by the minimum water loading.

2. In order to minimize the floor space used by the filter, the filter should be as tall as possible. The practical limitations are the height of the building, the head limits on the pump and the structural and stability considerations of the vessel.

3. A taller filter will have a longer flow path for the water. This means a more complete treatment of the water with each pass.

4. Taller filters will have higher specific water loadings. This means better flushing action, more turbulent water films and higher ammonium removal rates.

Trickling filters for industrial applications are sometimes 30 ft. tall. This is not practical for aquaculture systems. In general, trickling filters for aquaculture are between 4 and 10 ft. tall.

Submerged Bed Filters

Submerged Bed Filters are familiar to anyone who has owned an aquarium. An under gravel filter is a classic down flow submerged bed filter. Submerged bed filters have been used extensively for

small scale aquaculture and backyard water feature systems. These filters can be operated in up
flow, down flow or cross (horizontal) flow. The classic (old) systems consisted of gravel with an
under drain system. An improvement to these systems was the addition of air piping underneath.
The air was used to 'bump' the filter to dislodge solids that plugged the gravel and restore full flow.
There are numerous problems with these types of filters. Their large size, low void fraction, tendency to plug and extremely high weight make them expensive to build and maintain. In general,
these old gravel based systems are not suitable for modern aquaculture.

Modern submerged bed filters are very efficient, have low head loss and are very easy to build and
maintain. The key difference is the type of media and the water flow path. A modern submerged
filter uses structured media in a horizontal flow mode. This type of biofilter probably comes closer
to the ideal biofilter than any other type.

A typical installation would be configured similar to a raceway. The filter media is installed in a
long trough. The length of the flow path can vary based on the retention time required. By using
a relatively high velocity, it is possible to insure plug flow. This is a big advantage over well mixed
systems or systems with short retention times. If it is not possible to remove all of the BOD before
the biofilter, one will establish different zones in the filter. As nutrients are absorbed or removed
in the first sections of the filter, different types of organisms will establish dominance in the zones
where they enjoy optimum conditions. There are a variety of ways to configure a raceway type
system. Here are a few examples

TOP VIEW - RACEWAY TYPE SUBMERGED BIOFILTER

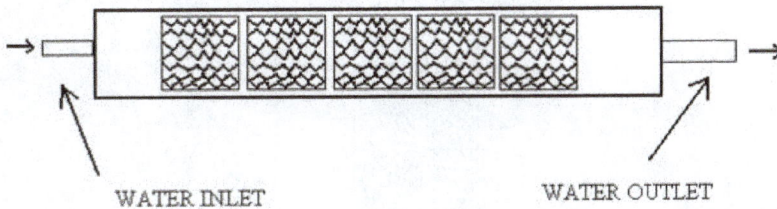

WATER INLET   WATER OUTLET

TOP VIEW - DOUBLE RACEWAY BIOFILTER

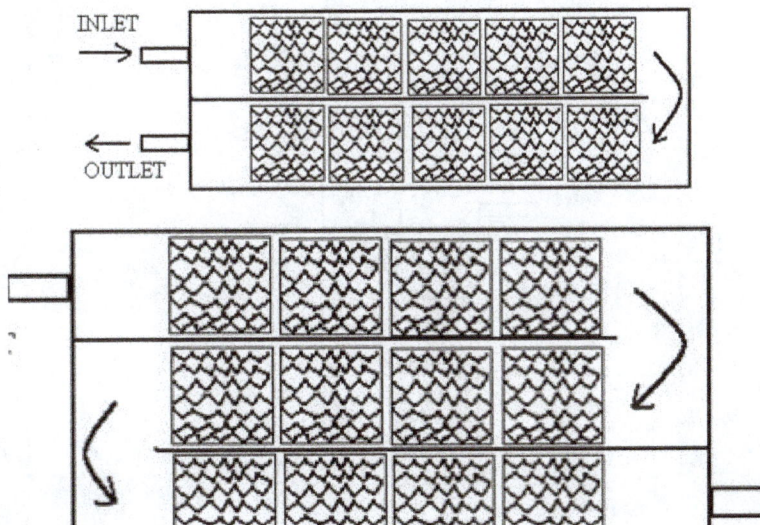

INLET

OUTLET

SIDE VIEW - FOLDED RACEWAY WITH VERTICAL FLOWS

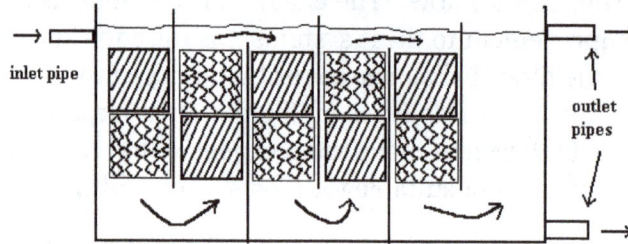

Submerged filters can operate with or without aeration. If the flow path is long and the nutrient loading is high, it is wise to have aeration in the filter. One of the easier methods is the traditional aeration system with large silica air stones.

Sometimes it is not possible to use a raceway type biofilter system. If existing tanks must be used, it might be easier to build a system with internal recirculation. The advantage of internal recirculation is that it increases the velocity of water past the media and adds oxygen to the water. Increasing the velocity helps insure a more even distribution of water throughout the filter media and reduces the possibility of dead zones that are not receiving nutrients and oxygen. It also helps to keep particles in suspension. Suspended solids tend to settle out in areas of low water velocity. This is a problem because accumulations of solids can become anaerobic and contribute to poor water quality. Here are a couple of examples of internal recirculation systems. The cone bottom tank is preferred over the flat bottom tank because any solids that settle out will be removed immediately.

SIDE VIEW - INTERNAL RECIRCULATION
TYPE BIOFILTER

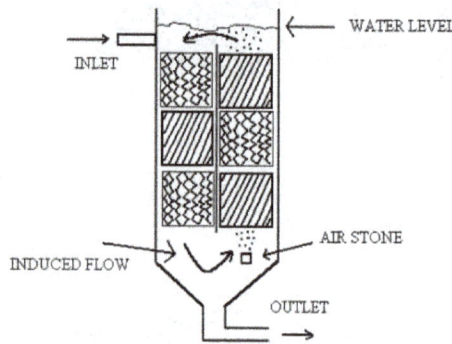

SIDE VIEW - INTERNAL RECIRCULATION
WITH FLAT BOTTOM VESSEL

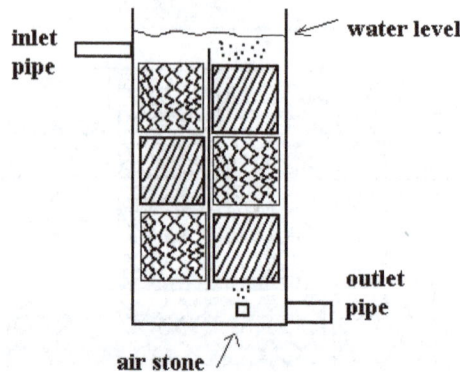

There is always the possibility to install the submerged biofilter media in the culture tank. This has the advantage of saving the cost a separate vessel and associated piping. The big disadvantage to this system is that it is difficult to remove the suspended solids before the water enters the biofilter. Because there are too many different configurations to draw them all, here is a brief description of a few of the possibilities.

1. Air lift the water into one end of a filter designed as a raceway and air lift it back into the culture tank at the other end.

2. Pump the water into a particulate filter such as a rotary drum and then flow through the biofilter.

3. Locate tubes or columns of packing throughout the culture tank and induce a flow through them with air stones.

4. Locate the filter media around the walls of the culture tank and induce a flow up through the media with air stones.

The number of possible configurations is limited only by one's imagination.

Submerged filters are excellent choices for small systems because they are very versatile. They can be located in a separate tank or in the culture tank. They can be horizontal flow, up flow or down flow. They can be aerated or not. The most important consideration for the design is the even distribution of water to the packing. It is very common for submerged filters to be designed as large, flat and thin sections of packing with water direction being up flow or down flow. There is typically no provision for distributing the water to all areas of the media. The length of the water path through the media is very short and the resistance to flow is very low. This is a recipe for disaster. The water flow will short circuit though a small section of the media and the rest of the biofilter will become anaerobic.

Ideally the flow path through a submerged filter should be as long as possible. A long thin raceway is the best. This type of biofilter is known as a long path, plug flow submerged filter. Another possible alternative is the use of aeration to induce a circulating flow around a tank. The goal should always be to provide sufficient velocity through the media to insure a fresh supply of oxygen and nutrients to the bugs on the surface of the media.

General Water Quality Maintenance Principles

Not all aquaculture applications have the same requirements for biofiltration. Not only do crops vary in their requirements but different farmers may grow the same crop under different conditions. The biofilter is only one of several components of the system used to maintain water quality. The functions that the biofilter must perform are determined by the presence and effectiveness of other components. Here are some other components and their effects on the system.

Aeration or Oxygenating Systems

If the fish don't have oxygen you are out of business no matter what else you do. Aeration is always the first step when increasing carrying capacity over an open, lightly loaded system. Mechanical surface aerators, subsurface air bubblers and pure oxygen injection is the typical progression in terms of technology and complexity. All aerobic biofilters require oxygen to operate. If the biofilter does not provide its own oxygen, it will be limited to the oxygen carried in with the water.

Particulate Filters

Once sufficient oxygen is provided, the next easiest way to improve water quality is to remove suspended solids. This is a more difficult task since particles come in all shapes, sizes and densities. Suspended solids consist primarily of uneaten food and feces which are slightly denser than water. Large particles, above 100 microns, will settle out quite easily. Particles above 50 microns can be filtered out with a screen. Particles below 10 microns are difficult to filter and are generally removed by some other means.

There are many different types of particulate filters that can remove suspended solids. They generally fall into three broad categories. The first type are settling basins, tube settlers, plate settlers, swirl separators and similar systems that allow the particles to drop out of the flowing stream by gravity. They are relatively simple devices and they work well on large particles. Settling systems generally have very low pump head requirements.

The second type are sand filters, sock filters, drum filters, disk filters, belt filters and similar systems that mechanically remove the particles from a flowing stream. These types of systems "screen" the particles. The size of particle removed is dependent on the size of the screen or sieve. Pump head requirements can vary from low to very high. Some biofilters such as bead filters claim to do both particulate filtration and biological filtration.

The third type of particulate filter is air floatation or fractionation. These are commonly known as protein skimmers. In this device, air is bubbled into a column and the fine particles become attached to the surface of bubbles. The resulting froth or foam is collected and removed from the system. These devices require a certain amount of surfactant type compounds in the water in order to work properly. Generally speaking, they work better in salt water than fresh water systems.

Although they are not typically designed for solids removal, some submerged biofilters will tend to collect fine particles due to the sticky nature of biofilms. This can be both a benefit and a maintenance problem. If the biofilter is not designed for easy cleaning, solids collection can represent a maintenance headache.

Removal of suspended solids is important since suspended solids comprise the majority of the BOD (Biological Oxygen Demand). The BOD not removed by the particulate filtration system must be removed by the biofilter before effective ammonia removal will occur. Thus the size of the biofilter is influenced by the effectiveness of the particulate filtration system.

The way that solids are removed is also important. The best systems remove solids quickly without degrading them in any way. If the solid particles are broken or reduced in size, it makes it easier for nutrients to dissolve into the water. These nutrients must then be removed by another part of the water treatment system or flushed out by water exchange. Time is also important because the longer solids are held in the system, the more degradation will occur. Floating bead filters are particularly bad in this regard since they hold the solids for long periods of time before backflushing.

Foam Fractionators

Foam fractionators are very useful but sometimes optional pieces of equipment. They are good at removing small particles (under 10 microns) and surface active compounds. They are sometimes referred to as protein skimmers. Since proteins are nitrogenous compounds that degrade into ammonia, foam fractionators can reduce the load on the biofilters. They are definitely useful in systems where water clarity is important. Foam fractionators also add oxygen to the water as a secondary benefit. Unfortunately, foam fractionators do not always work well in fresh water.

Ozone

Ozone is a powerful oxidizer and sterilant. It is potentially harmful to fish, humans and most living organisms. It is definitely harmful to biofilters. It is used to improve water clarity and reduce disease transmission. Ozone should never be used directly before a biofilter. If ozone is used upstream of a biofilter, there should be sufficient retention time after the injection point to insure that no ozone residual enters the biofilter.

UV light

Certain wavelengths of UV (Ultraviolet) light can be used as a sterilant. UV light is often used with ozone. UV light and ozone are complimentary and synergistic.

Carbon Dioxide Strippers

Build up of CO_2 can be a serious problem in a heavily loaded, intensive recirculating system using pure oxygen. The choice of biofilter has a direct influence on the degree to which CO_2 is a problem. In general, any biofilter other than a trickling filter or RBC will have a CO_2 problem when pure oxygen is used rather than compressed air for aeration. Building a CO_2 stripper is not a difficult task but it must be included in the overall design of the system.

In order to remove carbon dioxide, there must be a large interfacial area between air and water. The interfacial area can be increased through the use of subsurface aeration, mechanical surface aerators, spray systems or packed columns. Subsurface aeration is not very efficient and mechanical surface aerators are difficult to use in an intensive recirculating systems. Spray systems can be big energy users and they are not very efficient either. The best choice for intensive and space limited systems is the packed column. Packed columns can be either cross flow or counter flow systems. Packed columns for CO_2 stripping require fans to either force (push) air in or induce (pull) air through the packing.

Fish Stocking

Fish stocking is a fish management tool that works by releasing fish, usually bred in hatcheries, into the wild. Just like grocery items, it seems, the public has a demand for fish. Most often, the irate shoppers are replaced by angry anglers who want to be able to fish for sport. In other instances, environmentalists are the ones who want to restock empty waters to mitigate for past losses due to habitat disturbances or to restore historic populations. Sometimes, the customer is someone with a backyard pond, while at other times, the customer is a scientist who wants to conduct research.

The guidelines for stocking fish vary by location, with different countries and states having their own rules about which fish can be stocked where by whom. But whether it's a governmental office or a backyard enthusiast doing the stocking, careful planning is essential. After all, if a grocery store didn't plan its restocking schedule, the shelves might end up with no bread but hundreds of cans of peas. Imagine if two shoppers got into a brawl over that last loaf of bread, and you can start to imagine what might happen in an inappropriately planned fish stocking. While the shoppers may eventually settle the matter, the fish would die.

References

- What-is-aquaponics-and-how-does-it-work: permaculturenews.org, Retrieved 11 April 2018
- Aquaponics-components: projectfeed1010.com, Retrieved 14 June 2018
- Methods-of-aquaponics: aquaponics.com, Retrieved 17 April 2018
- Deep-water-culture-154: maximumyield.com, Retrieved 27 June 2018
- Bio-filter-designs-and-explanation: growfish.com, Retrieved 19 July 2018

Permissions

All chapters in this book are published with permission under the Creative Commons Attribution Share Alike License or equivalent. Every chapter published in this book has been scrutinized by our experts. Their significance has been extensively debated. The topics covered herein carry significant information for a comprehensive understanding. They may even be implemented as practical applications or may be referred to as a beginning point for further studies.

We would like to thank the editorial team for lending their expertise to make the book truly unique. They have played a crucial role in the development of this book. Without their invaluable contributions this book wouldn't have been possible. They have made vital efforts to compile up to date information on the varied aspects of this subject to make this book a valuable addition to the collection of many professionals and students.

This book was conceptualized with the vision of imparting up-to-date and integrated information in this field. To ensure the same, a matchless editorial board was set up. Every individual on the board went through rigorous rounds of assessment to prove their worth. After which they invested a large part of their time researching and compiling the most relevant data for our readers.

The editorial board has been involved in producing this book since its inception. They have spent rigorous hours researching and exploring the diverse topics which have resulted in the successful publishing of this book. They have passed on their knowledge of decades through this book. To expedite this challenging task, the publisher supported the team at every step. A small team of assistant editors was also appointed to further simplify the editing procedure and attain best results for the readers.

Apart from the editorial board, the designing team has also invested a significant amount of their time in understanding the subject and creating the most relevant covers. They scrutinized every image to scout for the most suitable representation of the subject and create an appropriate cover for the book.

The publishing team has been an ardent support to the editorial, designing and production team. Their endless efforts to recruit the best for this project, has resulted in the accomplishment of this book. They are a veteran in the field of academics and their pool of knowledge is as vast as their experience in printing. Their expertise and guidance has proved useful at every step. Their uncompromising quality standards have made this book an exceptional effort. Their encouragement from time to time has been an inspiration for everyone.

The publisher and the editorial board hope that this book will prove to be a valuable piece of knowledge for students, practitioners and scholars across the globe.

Index

www.ingramcontent.com/pod-product-compliance
Lightning Source LLC
Chambersburg PA
CBHW082026190326
41458CB00010B/3284